ML.NET Revealed

Simple Tools for Applying Machine Learning to Your Applications

Sudipta Mukherjee

Apress®

ML.NET Revealed: Simple Tools for Applying Machine Learning to Your Applications

Sudipta Mukherjee
Bangalore, India

ISBN-13 (pbk): 978-1-4842-6542-0
ISBN-13 (electronic): 978-1-4842-6543-7
https://doi.org/10.1007/978-1-4842-6543-7

Managing Director, Apress Media LLC: Welmoed Spahr
Acquisitions Editor: Joan Murray
Development Editor: Laura Berendson
Coordinating Editor: Jill Balzano

Cover image designed by Freepik (www.freepik.com)

Distributed to the book trade worldwide by Springer Science+Business Media LLC, 1 New York Plaza, Suite 4600, New York, NY 10004. Phone 1-800-SPRINGER, fax (201) 348-4505, e-mail orders-ny@springer-sbm.com, or visit www.springeronline.com. Apress Media, LLC is a California LLC and the sole member (owner) is Springer Science + Business Media Finance Inc (SSBM Finance Inc). SSBM Finance Inc is a **Delaware** corporation.

For information on translations, please e-mail booktranslations@springernature.com; for reprint, paperback, or audio rights, please e-mail bookpermissions@springernature.com.

Apress titles may be purchased in bulk for academic, corporate, or promotional use. eBook versions and licenses are also available for most titles. For more information, reference our Print and eBook Bulk Sales web page at http://www.apress.com/bulk-sales.

Any source code or other supplementary material referenced by the author in this book is available to readers on GitHub via the book's product page, located at www.apress.com/9781484265420. For more detailed information, please visit http://www.apress.com/source-code.

Printed on acid-free paper

For you Sohan, my boy!

Table of Contents

About the Author

Sudipta Mukherjee is an electronics engineer by education and a computer scientist by profession. He holds a degree in electronics and communication engineering. He is passionate about data structure, algorithms, text processing, natural language processing tools development, programming languages, and machine learning. He is the author of several technical books. He has presented at @FuConf and other developer events, and he lives in Bangalore with his wife and son. He can be reached on Twitter @samthecoder.

About the Technical Reviewer

Olia Gavrysh is a program manager, speaker, and writer. Currently, she works at Microsoft and focuses on .NET 5 and .NET desktop. Before, she was the PM for .NET machine learning framework called ML.NET from its creation to bringing the product to open source and releasing the preview. Her background is in applied mathematics and artificial intelligence. Before becoming a PM, she was a software developer working with .NET stack. She can be reached on Twitter @oliagavrysh.

Acknowledgments

A book like this one is not the fruit of a single person's labor. This is a team effort although that doesn't quite show up on the surface, where the name on the book's jacket reflects that of the author. So I take this opportunity to express my gratitude for people without whom the book wouldn't have hit the press at all.

Writing is hard. If you have written anything significant, you know that you shall suffer from writer's block, feeling completely blank in your head about what to write while funnily enough have a clear understanding of what would the pretty picture look like when the writing gets completed. In these situations, you need people who understand that situation and keep backing you up and wait for the content. I am not only fortunate but would say rather blessed to work with two individuals with a great deal of patience and grit. They are my acquisitions and coordinating editors *Joan Murray* and *Jill Balzano*. They constantly waited for my contents to appear and when I failed waited again. At moments I felt like I am walking on a slippery slope and wouldn't probably make it in any time at all, let alone soon. But miracles do happen when you have such wonderful people around you. Thanks a lot Joan and Jill. It was quite a ride and am surely looking forward to more in the future. Hopefully I shall disappoint you a little less in the future☺.

The next person I am indebted to is *Olia Gavrysh*. She is a program manager in the .NET team and previously managed the ML.NET team. She agreed to review the text when I approached her. This is very kind of her and she was very fast and accurate in providing eye-opening feedback that really improved the quality of the book and expanded my knowledge as well. When someone like herself, who has spent quite a lot of those initial days with the ML.NET team, comes forward and reviews the text, and when she approves what I have to say about ML.NET, it means a lot to me. I can't thank you enough Olia for doing this for the book. I seriously hope that we remain connected for future projects!

ACKNOWLEDGMENTS

Last but not the least at all, I want to thank my wife *Mou* for always keeping my morale high and standing by me whenever I needed it the most. Thank you sweetheart for all the love and sweat. I know I ruined many of your evenings and dinners for the writing, and I can't thank you enough for giving me space when I needed to focus on the book.

Finally, I want to thank the Almighty for this awesome opportunity and placing immensely kind and loving people in my life and for helping me keep dreaming. I have had enormous fun writing the book. I hope you shall love it in your journey to the wonderland of machine learning.

Introduction

Thanks for picking this book. This will introduce you to the wonderful world of machine learning via Microsoft's open source cross-platform framework ML.NET.

That means if you master this framework, you can write machine learning (a.k.a. ML) applications or applications that use ML and run it on all platforms (Windows, Linux, MacOS).

Here is a brief introduction to the chapters.

Chapter 1: Meet ML.NET (*Nothing is magical, but a few things seem so*)

This chapter introduces you to the ML.NET framework and gives a very brief overview of tasks that are possible via ML.NET.

Chapter 2: The Pipeline (*Great Machine Learning requires great plumbing*)

This chapter introduces you to the plumbing that needs to happen in order for your ML tasks to be successful.

Chapter 3: Handling Data *(Cleansing is engineering)*

Data come in different formats and mostly are messy when they are onboarded in a system. This chapter shows how to clean data using several transformations offered by ML.NET.

Chapter 4: Regressions (*How much will our dream home cost?*)

This chapter shows how to use regression algorithms to predict prices of things in the future.

Chapter 5: Classifications (*Helping computers tell chalk and cheese apart*)

Classifying one object from another (a.k.a. binary classification) and classifying many objects in different categories (a.k.a. multiclass classification) are two classic ML tasks that are solved using ML.NET in this chapter.

Chapter 6: Clustering (*Birds of a feather flock together*)

Grouping things automatically into different groups is called clustering, and this is a classic unsupervised learning algorithm. This chapter shows how to solve these problems using ML.NET framework.

Chapter 7: Sentiment Analysis (*Are you happy or not, that's the question!*)

Automatically detecting polarity (positive or negative) from phrases is really important business and is an active research area. This chapter shows how to do sentiment analysis using ML.NET and some other techniques that are yet to appear in ML.NET but will sure do soon.

Chapter 8: Product Recommendation (*You might be interested in this movie*)

Product recommendation boosts product sales, and this chapter shows how you can use popular techniques like collaborative filtering and matrix factorization using ML.NET for product recommendations.

Chapter 9: Anomaly Detection (*That doesn't look normal. Does it?*)

Detecting odd ones from a pool of products is key to the success of a manufacturing business at this time because it is inhuman to expect that human employees can monitor everything. That's where anomaly detection comes in to help. This chapter is dedicated to those algorithms and how to do those using ML.NET.

Chapter 10: Object Detection (*Can you spot the cat in the photo?*)

Detecting objects, faces from a photo or video frame is all the trend these days and has many applications. This chapter shows how to use ML.NET to do object detection using deep learning via ML.NET and ONNX.

CHAPTER 1

Meet ML.NET

Nothing is magical, but a few things seem so

Machine learning is nothing but a means of enabling the computer to have a sophisticated *sense of proximity* between several things. Let me elaborate that point for you with a few examples. Human vision is very advanced. So much so that we hardly realize what is going on in our brain when we recognize something. For example, do you think about the complex processes running in your brain when you read a handwritten note and recognize that is a letter "a"? Consider the pictures of the letter "a" in Figure 1-1.

Figure 1-1. *"a" written in multiple fonts*

We recognize each of these as the letter "a" because although they look different, they are within a *permissible range of* **proximity** from the "ideal" (if you will) "a" that we were taught in our childhood. Teaching a computer to recognize things is no different. We must provide the algorithm several examples with labels, and eventually the algorithm will start to spot similar things with better results. This approach is called *supervised learning* and will be explained in more detail in further chapters.

© Sudipta Mukherjee 2021
S. Mukherjee, *ML.NET Revealed*, https://doi.org/10.1007/978-1-4842-6543-7_1

Figure 1-2. *Different types of wooden shapes*

Another type of learning that we develop without realizing is the capability of segregating things (also known as *clustering*) without much input from outside. For example, if you present the shapes shown in Figure 1-2 to a toddler and tell them to determine how many different types of things are there, the answer will be 6. I urge you to look at the picture and determine the number yourself. The problem of this is you know the result, but how did you arrive at that is difficult to convey. This makes me remember this great quote (Figure 1-3).

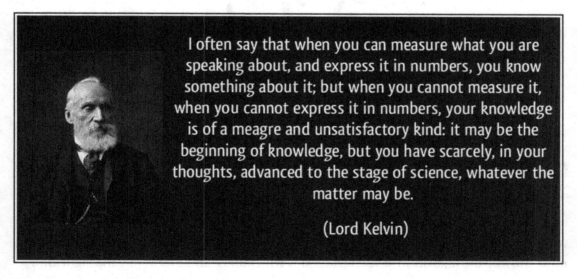

Figure 1-3. *Quotation of Lord Kelvin*

Throughout the book, we will consider more examples like this where the task will be to identify different types of things automatically without being told how many there are. The task they have in common is that these sorts of questions don't have a correct answer known ahead of time (e.g., how many different shapes are there). This is known as "unsupervised learning."

For the first group, you can think of it like a class with pupils and a teacher that is asking questions and telling the kids if they are correct or not. And that's why it is called "Supervised." In the second case, we don't know the answer – we don't have a supervisor.

There is another kind of learning that is reinforced by the experience of good and bad outcomes of the tasks performed. Do you remember how you learned to walk? Can you teach a baby or a robot to walk? We learn to walk because our brain had been continuously taking cues from the bad and good steps we took. Teaching a computer to do similar things is similar. All we must do is provide the computer with several opportunities to do mistakes and learn from the outcomes. Good outcomes will reinforce the belief of the algorithm that the steps taken were good, and bad outcomes will reinforce the fact that the steps taken were bad and therefore advisable to avoid. This type of learning is called "reinforcement learning" in machine learning literature. This is a little hard to follow along just by reading text. This is something to feel. I urge you to watch this video of a robotic arm throwing objects: `www.youtube.com/watch?v=JJ1Sgm9OByM`.

Abstraction matters

What is your favorite concept in object-oriented programming? Mine is *abstraction*. A good abstraction makes everything look easy. Achieving good abstraction over complex things/domains like machine learning, for example, is very hard because identifying which part would be a great choice for a building block is difficult at best and impossible at worst; but ML.NET does a great job striking a balance.

Note   As you know, this book is about ML.NET, Microsoft's new ML framework for .NET developers released in 2019. It allows developers to enhance their application with ML capabilities, but the best thing about it is that you don't need to learn data science and math to be able to use it.

ML.NET *democratizes* machine learning by bringing it to .NET developers who have been developing line-of-business applications for enterprises, web pages, applications, and what-have-you since ages and now facing the challenge to solve machine learning problems because enterprises have gargantuan amount of data and they want their existing staff to help them turn these data into actionable insights – fast. It's a tall order. Not an easy task at all, but a good framework like ML.NET can help.

ML.NET *encapsulates* machine learning algorithms such that most of the time using the algorithm *merely* becomes *calling* a function. This can seem to be an *oversimplification*, but this makes it *easy* for developers who don't really need to understand how the algorithm works internally, to consume the algorithm, thereby removing/reducing the barrier of entry – if you will, into the machine learning arena. Using an algorithm and assessing its performance based on some preset matrices is one thing, and understanding how the algorithm works internally is a completely different thing. For the most part, however, it is enough for developers to know how to use an algorithm and how to measure its performance for the task at hand, so that the parameters can be changed for optimization and they (developers) can do away with requiring to acquire the knowledge of really understanding what goes under the hood.

Consider the example of linear regression. Don't worry if you don't understand what linear regression is. For now, it is enough to know that it is a way to fit a few points to a given straight line so that predictions can be made about new input points.

$$y_i = \beta_0 + \beta_1 x_{i1} + \cdots + \beta_p x_{ip} + \varepsilon_i = \mathbf{x}_i^\mathsf{T} \boldsymbol{\beta} + \varepsilon_i, \qquad i = 1, \ldots, n,$$

Figure 1-4. *General equation for regression*

The preceding equation is a generic form of linear regression. There are however several varieties, and knowing all the details about them is beyond the reach of affordable time commitment that a .NET developer can spend keeping their day job. This is just an example of how complex machine learning models can be. But a good framework like ML.NET can save all the details except the ones that are absolutely required to tune the algorithm. Also, developers can learn how to use a framework if it is presented well.

For a second, compare these two things side by side.

$$y_i = \beta_0 + \beta_1 x_{i1} + \cdots + \beta_p x_{ip} + \varepsilon_i = \mathbf{x}_i^\mathsf{T} \beta + \varepsilon_i, \qquad i = 1, \ldots, n,$$

```
var model = pipeline.Fit(dataView);
```

The call of the **Fit** method on the right will look immediately known to .NET developers, and they seemingly don't need a mathematical background (tall claim but true) to understand what is happening behind the scene. Moreover, depending on the input, the **Fit** method can choose to use different regression algorithm and therefore be more efficient than the hard-coded model. Also these models will need to evolve, and it will be difficult for the average .NET developer (with all due respect to them) to keep up with all the advances, and therefore if hard-coded, sooner or later the model will be missing out on the enhancements made in Machine Learning theory by the scientists. However, if a framework like ML.NET is used, then these enhancements are expected to make their move in the framework, and without knowing, the developers will be able to take the benefits. Therefore, *abstraction matters*, and a good abstraction makes almost anything look very simple.

So while the actual equation on the left will appeal to mathematicians, the calls to the framework on the right will make developers happy, and that's the motivation of the framework. In a nutshell, the goal of ML.NET is "To make the cliché that Machine Learning is a niche."

The framework provides functionalities for all parts of the machine learning pipeline starting from data acquisition to model evaluation and cross-validation (checking how the algorithm did). The framework also encapsulates several feature engineering techniques in the form of generic methods that eases the process of data preparation in a really efficient and clean way.

What type of problems can be solved with ML.NET?

ML.NET supports "supervised learning" and "unsupervised learning" as of now (September 2020). This will surely change in the future, but this is a good start because many useful projects are relying on supervised /unsupervised learning.

The following is a list of some very common machine learning types of problems along with their class of problem domain:

- Supervised Learning
  - Regression (predicting real values)
    - Predicting prices of houses
    - Predicting temperature on a specific day
  - Classification

- Binary classification (telling chalk and cheese apart)

    - Telling cancer and non-cancer cells apart

    - Telling chalk and cheese apart

- Multiclass classification (autocategorization, more than one)

    - Identifying flowers by sizes of petals and sepals as their features

    - Tagging GitHub issues with corresponding labels

- Product recommendation

- Unsupervised Learning

    - Clustering (segmenting buyers in supermarket)

The Pipeline

The goal of most (or should I say all) of the machine learning activity is to come up with a model that maps a set of inputs to a predefined set of outputs as neatly as possible. Also, most of the time, data comes in messy ways unconsumable for machine learning algorithms.

Parts of ML.NET

ML.NET is built around a central type called "MLContext". Almost all operations performed using ML.NET use one or the other part of MLContext type. It is almost semantically like that of a DataContext. This class has several functionalities offered through different members:

- Data Loading

- Data Transformation

- Prediction

- Measuring Accuracy

Data Loading

ML.NET offers functionalities to load data from several different formats through several static methods. Here, some of them are shown.

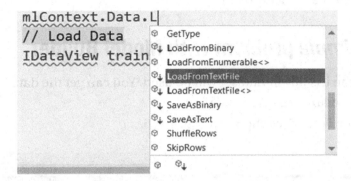

Figure 1-5. *Several ways to load/create data for machine learning problems*

As you can see, there are methods to load data from a few different formats. This is an incremental list. Soon many other data loaders may be supported.

Data Transformation

MLContext offers data transformation functionalities via Transforms property of MLContext class. Transforms is of type TransformCatalog. All these transformations are provided as extension method. So, if required we can also build our own transformation.

Prediction

Using the MLContext and related classes in ML.NET, we need to build a pipeline which represents the machine learning pipeline with components to do data loading, transformation, and prediction. A pipeline is used to create what is known as an estimator. The estimator is used to generate the model.

Measuring Accuracy

The performance of a machine learning model is measured by so many ways. ML.NET offers several metrics (depending on the task performed) to measure this performance, like confusion matrix and cross-validation accuracy.

Introduction to Model Builder (Automatic ML)

ML.NET is designed with absolute beginners of machine learning in mind. So apart from several APIs to create a custom ML model, ML.Net also offers a fantastic UI-driven utility that helps beginners locate the best algorithm for solving a machine learning problem.

It runs several models and keeps track of accuracies and time taken to complete the training. This tool is very helpful in locating the algorithms.

In the next few sections, just a quick sample shows how to use this utility to solve a real-world machine learning problem.

Solving a *simple* problem with Model Builder

In this example, the Iris flower dataset will be used. You can get the data from https://archive.ics.uci.edu/ml/datasets/iris.

Go to the "Data Folder" as shown in Figure 1-6.

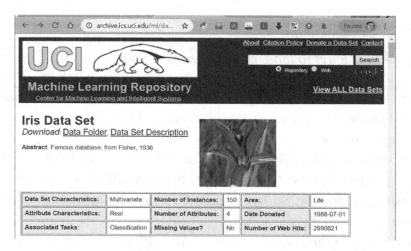

Figure 1-6. *Locating data for multiclass classification problem (Iris flower)*

Once you clicked the link for Data Folder, you shall see this (Figure 1-7).

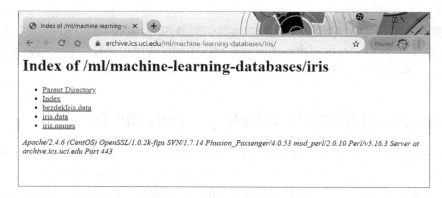

Figure 1-7. *Locating the data in UCI database for machine learning*

From here, click "iris.data" to download this file. This file will look like this (the first few rows and columns are shown in Figure 1-8).

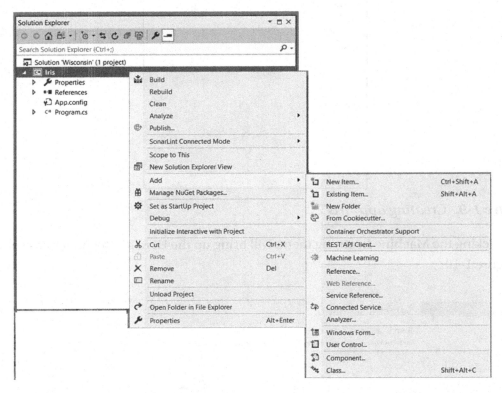

Figure 1-8. *Menu to add "Machine Learning" using Model Builder in Visual Studio*

The first column is the ID of the patient, and the second column denotes the diagnosis (either M for Malignant cancer or B for Benign cancer). The rest of the columns denote several values for several test results. The actual names for each of these columns are not important.

If you don't already have the Model Builder, then you can download it from the following link:

`https://marketplace.visualstudio.com/items?itemName=MLNET.07`

After installation, create a console app called "Iris", and the project will look like Figure 1-9.

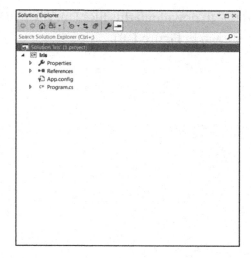

Figure 1-9. *Creating app "Iris"*

Clicking the Machine Learning menu will bring up the UI of Model Builder as shown in Figure 1-10.

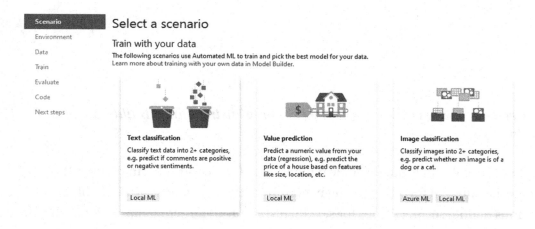

Figure 1-10. *Model Builder wizard interface (select "Text Classification")*

The first step is to understand that identifying a flower is a case of multiclass classification problem because we have three different classes of iris: versicolor, sentosa, and virginica. In other words, it is more like GitHub issue classification problem. So to identify flowers from data, we need to select that scenario. Once the button "Issue Classification" is clicked, the wizard presents the next screen to load the data and for setting parameters.

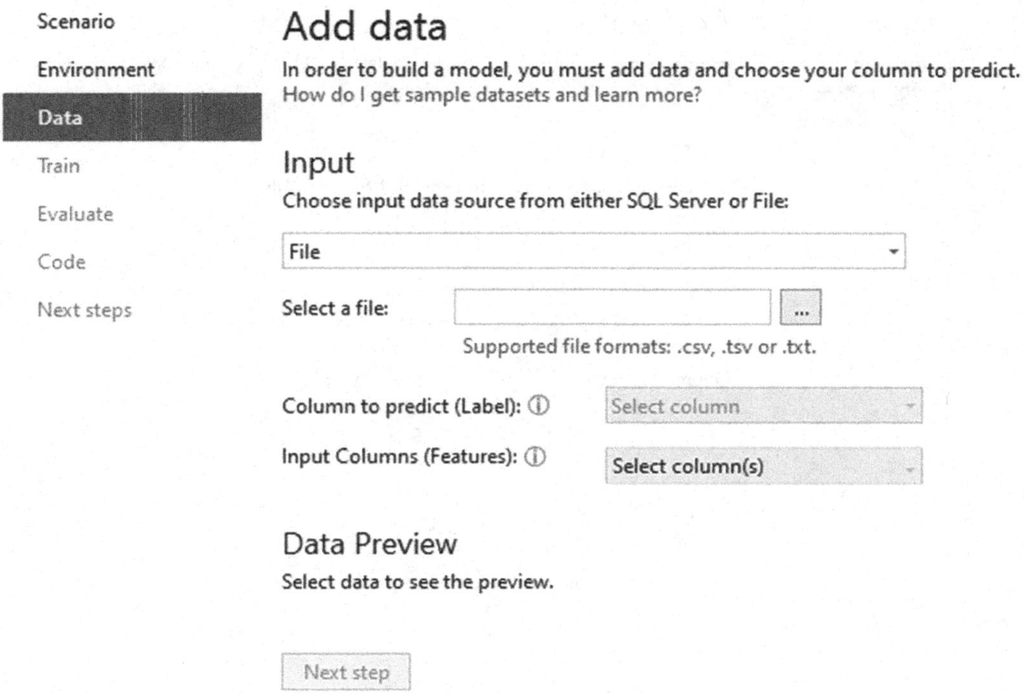

Scenario

Environment

Data

Train

Evaluate

Code

Next steps

Add data

In order to build a model, you must add data and choose your column to predict. How do I get sample datasets and learn more?

Input

Choose input data source from either SQL Server or File:

| File   ▾ |

Select a file: [   ] [...]

Supported file formats: .csv, .tsv or .txt.

Column to predict (Label): ⓘ   [Select column   ▾]

Input Columns (Features): ⓘ   [Select column(s)   ▾]

Data Preview

Select data to see the preview.

[Next step]

Figure 1-11. *Model Builder wizard interface for loading training data*

Once the input data is loaded, the wizard shows the data in preview as shown in Figure 1-12. Be aware that you need to rename the iris.data to iris.csv; otherwise, you won't be able to open it in the Model Builder.

Scenario

Environment

Data

Train

Evaluate

Code

Next steps

Add data

In order to build a model, you must add data and choose your column to predict.
How do I get sample datasets and learn more?

Input

Choose input data source from either SQL Server or File:

File	▾

Select a file: D:\iris.csv   `...`

Supported file formats: .csv, .tsv or .txt.

Column to predict (Label): ⓘ | Select column | ▾ |

Input Columns (Features): ⓘ | Select column(s) | ▾ |

Data Preview

10 of 151 rows and 5 of 5 columns.

sepallength	sepalwidth	petallength	petalwidth	variety
5.1	3.5	1.4	.2	Setosa
4.9	3	1.4	.2	Setosa
4.7	3.2	1.3	.2	Setosa
4.6	3.1	1.5	.2	Setosa
5	3.6	1.4	.2	Setosa
5.4	3.9	1.7	.4	Setosa
4.6	3.4	1.4	.3	Setosa
5	3.4	1.5	.2	Setosa
4.4	2.9	1.4	.2	Setosa
4.9	3.1	1.5	.1	Setosa

Figure 1-12. *Model Builder wizard interface for loading data and previewing and setup*

The next step is to tell the wizard which column we want to use for the Labeling, the column we want to predict. In this case, we shall need to use the "variety" column. Once this is done, the wizard marks the variety column as "Label" as shown in the data preview. The remaining columns are used for predicting the label. However, we can choose the columns to be used by selecting/unselecting the check boxes that appear before each column name.

Scenario

Environment

Data

Train

Evaluate

Code

Next steps

Add data

In order to build a model, you must add data and choose your column to predict.
How do I get sample datasets and learn more?

Input

Choose input data source from either SQL Server or File:

File	▼

Select a file: D:\iris.csv   [...]

Supported file formats: .csv, .tsv or .txt.

Column to predict (Label): ⓘ   variety   ▼

Input Columns (Features): ⓘ   4 of 4 columns selected   ▼

Data Preview

10 of 151 rows and 4 of 5 columns.

variety (Label)	sepallength	sepalwidth	petallength	petalwidth
Setosa	5.1	3.5	1.4	.2
Setosa	4.9	3	1.4	.2
Setosa	4.7	3.2	1.3	.2
Setosa	4.6	3.1	1.5	.2
Setosa	5	3.6	1.4	.2
Setosa	5.4	3.9	1.7	.4
Setosa	4.6	3.4	1.4	.3
Setosa	5	3.4	1.5	.2
Setosa	4.4	2.9	1.4	.2
Setosa	4.9	3.1	1.5	.1

[Next step]

Figure 1-13. *Setting up which field is to be predicted (Label)*

The next step is to train to obtain a model. The more time is given for the wizard to train, the better. This is because the wizard under the hood uses automatic machine learning to figure out which model is the best.

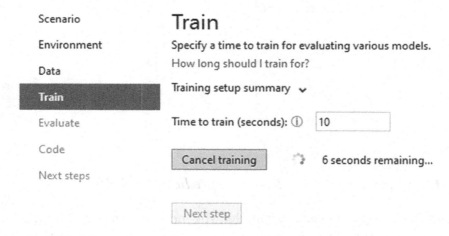

Figure 1-14. *Model Builder Wizard: Train Model interface*

The default training time given by the wizard is 10 seconds. Ten seconds is enough for datasets less than 10Mb. But it is strongly recommended that we use at least 90 seconds for the training for really small dataset.

Figure 1-15. *Showing progress of Model Builder Training Wizard*

Once the time is set to train the model, clicking "Start Training" will start the training, and the performance of the algorithms tried so far will be listed as shown.

Once the training is complete, the results of the algorithms tried can be viewed from the Evaluate tab.

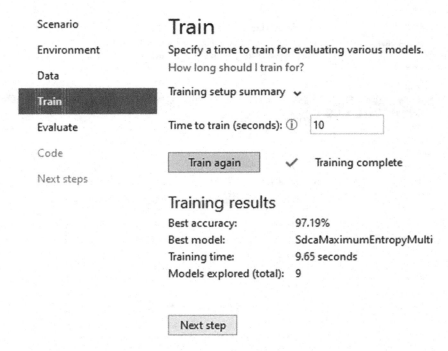

Figure 1-16. *Model Builder Training Model completed*

The Evaluate tab shows the details of the performance of the algorithms.

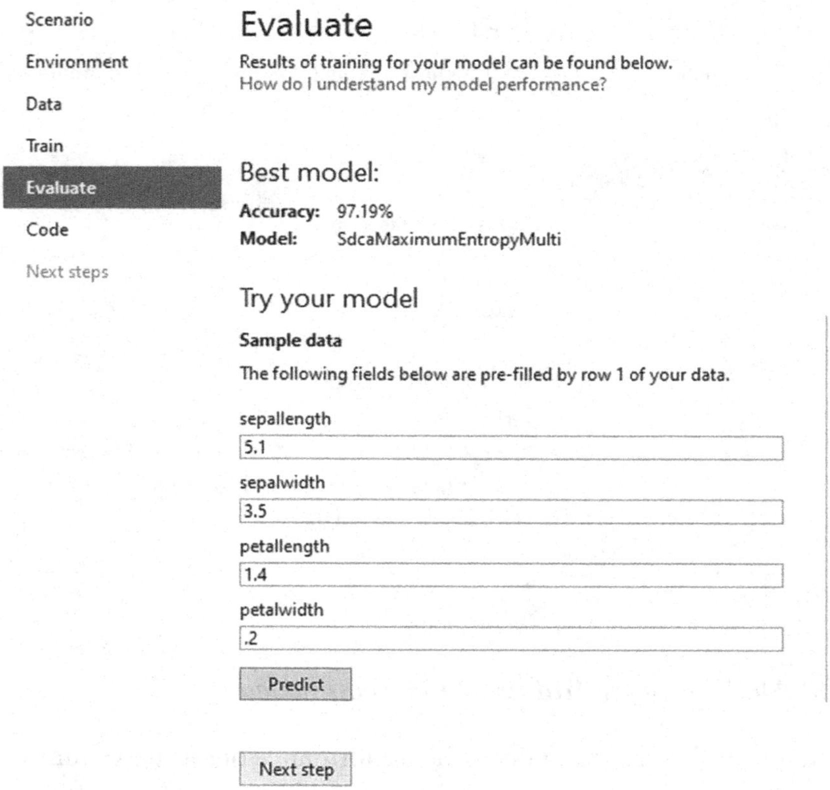

Scenario

Environment

Data

Train

Evaluate

Code

Next steps

Evaluate

Results of training for your model can be found below.
How do I understand my model performance?

Best model:

Accuracy: 97.19%
Model: SdcaMaximumEntropyMulti

Try your model

Sample data

The following fields below are pre-filled by row 1 of your data.

sepallength

5.1

sepalwidth

3.5

petallength

1.4

petalwidth

.2

Predict

Next step

Figure 1-17. *Model Builder Wizard Model Evaluate interface*

Interestingly, the Evaluate tab presents a nice interface to try out the model. This interface is autogenerated from the fields that were used to generate the model.

Scenario

Environment

Data

Train

Evaluate

Code

Next steps

Evaluate

Results of training for your model can be found below.
How do I understand my model performance?

Best model:

Accuracy: 97.19%
Model: SdcaMaximumEntropyMulti

Try your model

Sample data   **Results**

The following fields below are pre-filled by row 1 of your data.   Setosa   100%

sepallength   Versicolor   < 1%

5.1   Virginica   < 1%

sepalwidth

3.5

petallength

1.4

petalwidth

.2

Predict

Figure 1-18. *Showing on-the-fly generated UI for testing the prescribed model by Model Builder*

The last step is to add the code for the generated model in the solution.

Figure 1-19. *Model Builder Wizard interface to add generated projects to solution*

By clicking the "Add to solution", the projects generated can be added to the solution. After the projects are added, the solution explorer will show these new projects as shown in Figure 1-20.

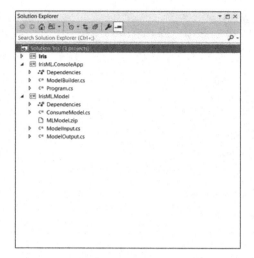

Figure 1-20. *Showing generated projects added to solution*

Walkthrough of the generated code

The Model Builder Wizard generates the ModelInput and ModelOutput class.

```
// This file was autogenerated by ML.NET Model Builder.
using Microsoft.ML.Data;
namespace IrisML.Model
{
    public class ModelInput
    {
        [ColumnName("sepallength"), LoadColumn(0)]
        public float Sepallength { get; set; }
        [ColumnName("sepalwidth"), LoadColumn(1)]
        public float Sepalwidth { get; set; }
        [ColumnName("petallength"), LoadColumn(2)]
        public float Petallength { get; set; }
        [ColumnName("petalwidth"), LoadColumn(3)]
        public float Petalwidth { get; set; }
        [ColumnName("variety"), LoadColumn(4)]
        public string Variety { get; set; }
    }
}
```

ModelInput class represents each row of the input dataset. The index of the column name is taken as the value of the LoadColumn attribute, and the name of the column is taken as the value of the ColumnName attribute. Here is one row taken from the input irisi. csv file. As you can see, the column indexing starts from 0.

```
sepallength,sepalwidth,petallength,petalwidth,variety
5.1,3.5,1.4,.2,Setosa
```

```
public class ModelOutput
{
    // ColumnName attribute is used to change the column name from
    // its default value, which is the name of the field.
    [ColumnName("PredictedLabel")]
    public String Prediction { get; set; }
    public float[] Score { get; set; }
}
```

ModelOutput class represents the prediction result. The score represents the score for all possible classes. Values of this Score array are shown on the test UI as percentage of confidence as shown in Figure 1-21.

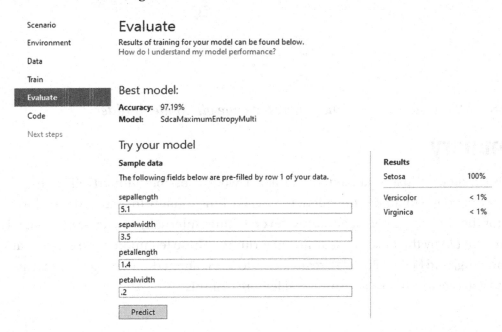

Figure 1-21. *Showing on-the-fly generated UI to test the prescribed model by Model Builder*

This means that for these set of test data, the model had 86% confidence that the flower is a **Setosa**, 14% confidence that it is a **Versicolor**, and less than 1% (or negligible) confidence that it is a **Virginica**. In code, these data can be represented as 0.86, 0.14, and 0.0034 as the elements of the Score array.

This line uses the Predict method of the ConsumeModel class (which is OK to remain as a black box for now) to return the output of the trained model as a ModelOutput instance.

```
// Make a single prediction on the sample data and print results
ModelOutput predictionResult = ConsumeModel.Predict(sampleData);
```

To make sure that the model code can't use the variety at all to predict (which will be obvious after looking at the code that it doesn't) the variety of the flower, "Unknown" is set to it. After that if you set the **IrisML.ConsoleApp** as the startup project and run the program, by putting a breakpoint as shown, you shall see a similar result (Figure 1-22).

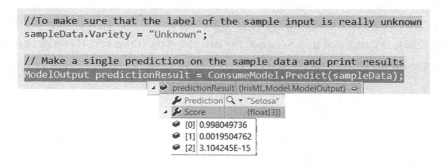

Figure 1-22. *Debug view of the generated code via Model Builder*

Summary

In this chapter, we just scratched the surface of what's possible with ML.NET. The framework does a lot under the hood. It provides functionality to load and transform data for the learning. Then, it also does several things internally to prepare the data that is consumable by the machine learning algorithms. These techniques are called feature engineering, and `Model Builder` really helps to learn about different algorithms and how to use these and how to use several features for this.

The framework is very modular and open for extension. More generally said, there are several extensions in the framework itself that build around core types.

In the next chapter, we shall use ML.NET to classification problems. You shall learn how to pose a classification problem as a binary or multiclass classification problems and how to use ML.NET to solve those.

CHAPTER 2

The Pipeline

Great Machine Learning requires great plumbing

Introduction

The goal of all machine learning (ML) activity is to turn raw data into some prediction or classification or insight. Raw data appears on the left or at the beginning of this pipeline, and on the right or at the end comes the insight/prediction/classification and so on. Although each machine learning task will require a different pipeline, the basic structure or the building blocks remain the same. ML.NET offers several types/interfaces to make the creation of this pipeline easier. A broad understanding of these concepts will help you understand how ML.NET works under the hood.

Objective of this chapter

After finishing this chapter, you shall be able to identify different building blocks of a machine learning pipeline and see that all ML.NET pipelines are essentially similar in nature although their purpose or the actual body is very different from one another. You shall learn to identify and tune all parts of all such machine learning pipelines.

The parts of the pipeline (in ML.NET)

- The context
- Data loaders
- Transformers
- Trainers

© Sudipta Mukherjee 2021
S. Mukherjee, *ML.NET Revealed*, https://doi.org/10.1007/978-1-4842-6543-7_2

Every machine learning operation in ML.NET is started by creating a machine learning context. The context is conceptually like the starting of the pipeline. It provides a way to create every part of the pipeline. The context is encapsulated in MLContext type.

The type has several properties to offer capabilities to start a specific machine learning task. At the beginning of each ML task in ML.NET, we must create a context object as shown in Listing 2-1.

Listing 2-1. Instantiating MLContext is easy

```
MLContext mlContext = new MLContext(seed: 1);
```

If seed is set, MLContext becomes deterministic and the same random numbers will be generated every time you run your app, so results will be repeatable across different runs. It can be helpful, if you are repeating a tutorial sample and want to get the same results. If you don't set seed, MLContext will use random numbers generator, and results will be slightly different for each run (for operation that use random numbers, not all of them do). In real life, I recommend keep random components nondeterministic, which means not setting seed.

Different kinds of machine learning activities are based on the pipelines created from MLContext properties. Here is the MLContext class.

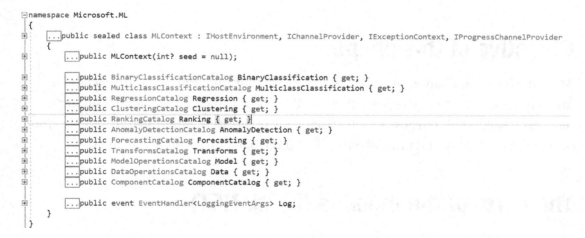

```
namespace Microsoft.ML
{
    public sealed class MLContext : IHostEnvironment, IChannelProvider, IExceptionContext, IProgressChannelProvider
    {
        public MLContext(int? seed = null);

        public BinaryClassificationCatalog BinaryClassification { get; }
        public MulticlassClassificationCatalog MulticlassClassification { get; }
        public RegressionCatalog Regression { get; }
        public ClusteringCatalog Clustering { get; }
        public RankingCatalog Ranking { get; }
        public AnomalyDetectionCatalog AnomalyDetection { get; }
        public ForecastingCatalog Forecasting { get; }
        public TransformsCatalog Transforms { get; }
        public ModelOperationsCatalog Model { get; }
        public DataOperationsCatalog Data { get; }
        public ComponentCatalog ComponentCatalog { get; }

        public event EventHandler<LoggingEventArgs> Log;
    }
}
```

Figure 2-1. *The overall definition of the MLContext class*

As mentioned earlier, MLContext acts as the root of the machine learning pipeline. Table 2-1 shows how different properties of this class are used for different types of machine learning problems.

Table 2-1. *Showing meaning of different properties of MLContext class*

Type of machine learning	Start the pipeline as
Binary Classification	`mlContext.BinaryClassification`
Multiclass Classification	`mlContext.MultiClassClassification`
Regression	`mlContext.Regression`
Clustering	`mlContext.Clustering`
Anomaly Detection	`mlContext.AnomalyDetection`
Forecasting (Time Series Data)	`mlContext.Forecasting`

`mlContext` is an object of the `MLContext` class in the preceding table.

Data loaders

Figure 2-2. *Representation of training data as bunch of files to process*

Data comes in several formats and sometimes it resides in memory collection. ML.NET offers features to load data from multiple sources easily. All these loaders can be accessed via mlContext.Data property as shown in Figure 2-3.

Figure 2-3. *Showing IntelliSense availability on MLContext instance*

Data in the pipeline travels inside IDataView type, introduced to .NET specifically for ML.NET. This is the input and output of Query Operators (Transforms). This is the fundamental data pipeline type, comparable to IEnumerable<T> for LINQ. This interface is required to be able to seamlessly integrate several data loading capabilities and for integrating with other machine learning frameworks.

Table 2-2 shows which function to be used to load data.

Table 2-2. *Showing several ways to load data*

Data Type/Purpose	Function to Load Data
Binary	LoadFromBinary
Data is in memory collection	LoadFromEnumerable
Loading data from text file as an IDataView	LoadFromTextFile
Loading data from text file in a strongly typed manner. The type of the data to be loaded is passed as the generic attribute	LoadFromTextFile<T>
Loading data from databases	mlContext.CreateDatabaseLoader<T>(). Load(...)

This is how loading from file looks like. So, if you have a CSV file, first create a ModelInput class representing each row of the CSV. Then, use this class as a generic parameter in LoadFromTextFile<T> as shown in the following example.

Listing 2-2. Loading training data to IDataView

```
IDataView trainingDataView =
            mlContext.Data.LoadFromTextFile<ModelInput>
        (path: TRAIN_DATA_FILEPATH,
         hasHeader: true,
         separatorChar: ',',
         allowQuoting: true,
         allowSparse: false);
```

If you don't like typing this class by hand every time you create an ML model, you can use the script (Listing 2-3) that automates this work for you.

The following C# script takes a CSV file and emits the ModelInput class.

Listing 2-3. Script to automatically generate code to load data

```
string csvFile = @"C:\MLDOTNET\iris.csv";
var columns = File.ReadLines(csvFile)
                .Take(1)
                .First()
                .Split(new char[]{','});
var firstLine = File.ReadLines(csvFile)
                .Skip(1)
                .Take(1)
                .First()
                .Split(new char[] { ','});
StringBuilder propertyBuilder = new StringBuilder();

for (int i = 0; i < columns.Length; i++)
{
    string column = columns[i];
    propertyBuilder.AppendLine($"[ColumnName(\"{column},LoadColumn({i})]");

    if(firstLine.ElementAt(i).ToCharArray()
        .All(m => Char.IsDigit(m) || m == '.'))
    {
```

```
            propertyBuilder
                .AppendLine($"public float {column.Substring(0, 1).ToUpper()
                        + column.Substring(1)}");
        }
        else
        {
            propertyBuilder.AppendLine($"public string
            {column.Substring(0,1).ToUpper() + column.Substring(1)}");
        }
        propertyBuilder.AppendLine("{ get; set;}");
}

string classCode = @"public class ModelInput " + Environment.NewLine
                + "{" + Environment.NewLine + propertyBuilder.ToString()
                        + Environment.NewLine + "}";
Console.WriteLine(classCode);
```

For the following CSV file, (first two rows of the data)

```
sepallength,sepalwidth,petallength,petalwidth,variety
5.1,3.5,1.4,.2,Setosa
4.9,3,1.4,.2,Setosa
```

Figure 2-4. *Showing the first couple of rows of the Iris training dataset*

It generates the following class. This script will save you countless hours typing your way to create the ModelInput to be just able to load the data and start your experiments.

Listing 2-4. ModelInput for the Iris dataset

```
public class ModelInput
{
    [ColumnName("sepallength"), LoadColumn(0)]
     public float Sepallength { get; set; }
    [ColumnName("sepalwidth"), LoadColumn(1)]
     public float Sepalwidth  { get; set; }
    [ColumnName("petallength"), LoadColumn(2)]
     public float Petallength { get; set; }
    [ColumnName("petalwidth"), LoadColumn(3)]
```

```
    public float Petalwidth  { get; set; }
    [ColumnName("variety"), LoadColumn(4)]
    public string Variety  { get; set; }
}
```

LoadColumn attribute specifies your properties' column indices and is required only when you load the data from file.

MLContext.Data also offers functionality to filter and shuffle data too apart from loading data from multiple sources.

Loading data from text files is a very common activity, and ML.NET is well equipped with it. It provides a couple of generic ways to load data from a text file, with or without headers.

The following code reads data from a Tab-separated file without headers where there are three numeric columns. Tab ('\t') is the default value of the separatorChar parameter.

Listing 2-5. Creating a TextLoader

```
var loader =
mlContext.Data.CreateTextLoader(
columns: new[]
{
  new TextLoader.Column("Feature1", DataKind.Single, 0)
  new TextLoader.Column("Feature2", DataKind.Single, 1)
  new TextLoader.Column("Feature3", DataKind.String, 2)

},
hasHeader: false
);
```

If you want to read a CSV without header, then you shall have to mention the separatorChar is "," as highlighted Listing 2-6.

Listing 2-6. Creating another custom TextLoader

```
var loader =
mlContext.Data.CreateTextLoader(
columns: new[]
{
```

```
  new TextLoader.Column("Feature1", DataKind.Single, 0)
  new TextLoader.Column("Feature2", DataKind.Single, 1)
  new TextLoader.Column("Feature3", DataKind.String, 2)
},
separatorChar: ',',
hasHeader: false
);
```

Now to read the data, you shall have to call the Load method on the loader just created like `loader.Load(<path_to_file>)`.

Since training data can sometimes be present in multiple files, it is required to create a loader and pass in the file paths as parameters.

Figure 2-5. *Showing overload of Load to read from multiple files*

Since data can be quite messy when it comes along, filtering could prove to be very useful to provide some initial cleansing that the data needs. `MLContext.Data` provides the following filtering capabilities.

Figure 2-6. *Showing filtering capabilities on IntelliSense*

Table 2-3. *Methods to perform filtering based on different criteria*

Filtering	What it does
FilterRowsByColumn	Filters data based on value ranges of a given column
FilterRowsByKeyColumnFraction	Filters rows by the value of a KeyDataViewType column
FilterRowsByMissingValues	Filters the data by dropping rows where any of the column in the passed list of columns have a missing value

Loading data from databases

To load data from a database into a IDataView, the following steps are required.

Listing 2-7. Loading training data from a database

```
//Name of the provider has to be given as "System.Data.SqlClient"
DbProviderFactory factory =
                DbProviderFactories.GetFactory("System.Data.
                SqlClient");
DatabaseSource = new DatabaseSource(factory,
                        "<connection string>",
                        "select * from IRIS");

IDataView trainingView = mlContext.Data.CreateDatabaseLoader<ModelInput>()
                                        .Load(databaseSource);
```

Transformers

As a caterpillar transforms into a butterfly, data must be transformed before using those in machine learning algorithms/models. Because if the data is fed directly to machine learning algorithms without proper prior cleaning, scaling, and normalizing, it will leave the algorithm confused, and the output of it will be biased, if not totally off, and that is absolutely unacceptable.

Figure 2-7. Symbolic representation of data transformation for machine learning

ML.NET offers several transformers to transform the data from messy to clean. Clean data means data that is free from any distortion and probably looks a lot like generated data in terms of clarity. Clean data don't have missing values, values out of permissible range for numeric columns, or impossible values for enumerations. For example, a

31

messy dataset for an insurance survey might have negative values for the age column
or impossible age values like 923 and so on. Similarly for the gender column of the
customer, the data can be outside the permissible enumeration like {M,F}. Clean data
is very important, and ML.NET offers several ways to clean the data via stages called
transformers. These transformations can be glued to one another to create a chain of
transformations or a pipeline if you will. The following is an example of such a pipeline
transformation.

Listing 2-8. Gluing transformations with Append

```
var pipeLine = context.Transforms.NormalizeMinMax("crim","crim")
            .Append(context.Transforms.NormalizeMinMax("zn","zn"))
            .Append(context.Transforms.NormalizeMinMax("indus",
            "indus"))
            .Append(context.Transforms.NormalizeMinMax("indus",
            "chas"))
            .Append(context.Transforms.NormalizeMinMax("indus", "nox"))
            .Append(context.Transforms.NormalizeMinMax("indus", "rm"))
            .Append(context.Transforms.NormalizeMinMax("indus", "age"))
            .Append(context.Transforms.Concatenate("Features",
            "crim", "zn", "indus", "chas", "nox", "rm", "age"))
```

Don't worry too much about the actual transformations. These will be explained
later in the book. For now, please note how transformations are glued to one another by
Append() method.

The first transformation `context.Transforms.NormalizeMinMax("crim","crim")`
returns a `NormalizingEstimator` as shown in this Listing 2-8.

The first Append() call on the pipeline earlier is an extension method created
on `NormalizingEstimator`. This extension method returns `EstimatorChain<N`
`ormalizingTransformer>`. Append() is also defined as an extension method on
`EstimatorChain<T>()`.

ML.NET heavily relies on extension methods to make things glue together nicely. In
the next chapter, you shall learn about several transformations in detail.

Trainers

Good trainers train the body; great ones train the mind. And finding a great trainer is hard at best, impossible at worst.

Figure 2-8. *Symbolic representation of a trainer*

The body is like the infrastructure of a machine learning system and the mind is like the actual model. If you have a great model and bad infrastructure, then it is bad, but if you have great infrastructure but bad model, it is worse, because in this setting you won't be able to use the full potential of the infrastructure, just as an untrained mind in a solid physique does.

ML.NET provides several trainers to train for different machine learning needs.

For each algorithm/task combination, ML.NET provides a component that executes the training algorithm and does the interpretation. These components are called trainers. For example, the `SdcaRegressionTrainer` uses the `StochasticDualCoordinatedAscent` algorithm applied to the Regression task.

All these trainers are in their respective types. For example, the binary classification trainers are located at `mlContext.BinaryClassification.Trainers`.

You shall learn more about the trainers later in the respective chapters of the book. However, it will be enough to know now that a trainer is added as the last step in the machine learning pipeline. Consider the example shown in Listing 2-9.

Listing 2-9. Adding trainer to the pipeline

```
var pipeLine = context.Transforms.NormalizeMinMax("crim","crim")
    .Append(context.Transforms.NormalizeMinMax("zn","zn"))
    .Append(context.Transforms.NormalizeMinMax("indus","indus"))
    .Append(context.Transforms.NormalizeMinMax("indus", "chas"))
    .Append(context.Transforms.NormalizeMinMax("indus", "nox"))
    .Append(context.Transforms.NormalizeMinMax("indus", "rm"))
    .Append(context.Transforms.NormalizeMinMax("indus", "age"))
    .Append(context.Transforms.Concatenate("Features",
    "crim", "zn", "indus", "chas", "nox", "rm", "age"))

    .Append(context.Regression.Trainers.OnlineGradientDescent(labelColumnName:"medv", featureColumnName:"Features",
    lossFunction: null, learningRate: 0.24f,decreaseLearningRate : true));

var model = pipeLine.Fit(trainingDataView);


var predEngine = context.Model.CreatePredictionEngine<BostonHouse, BostonHousePrice>(model);
```

The last `Append()` call creates an `EstimatorChain<RegressionPredictionTransformer>`.

So basically, a trainer is an algorithm that takes a data view and provides a model that can be applied to create a model, which in turn can be used to predict future values.

In Listing 2-9, the last Append call adds the trainer:

```
.Append(context.Regression.Trainers.OnlineGradientDescent(labelColumnName:"
medv",lossFunction:null, learningRate:0.24f, decreasingLearningRate:true));
```

This line sets hyperparameters (parameters that help the trainer to converge and are set before the iterative process begins) for the selected trainer OnlineGradientDescent.

Don't worry too much about the exact working of this code. This is to illustrate how common interface allows the trainer to be plugged into the pipeline as the last step. In later chapters, you shall learn how to pick a trainer for your machine learning task and how to evaluate their performances.

Model Builder (the wizard)

Figure 2-9. *Symbolic representation of the Model Builder*

It can be quite challenging for newbies and practitioners who are new to ML.NET to locate the right method for transformations or training. To address this problem, Microsoft created a wizard called `Model Builder`. This wizard can do all the data science decision-making part for you, suggesting you the best trainer with the best parameters for your particular case. As an input, you provide your dataset and the task (for instance, predict a house price), and as a result, you will see all the trainers that the wizard tried for your task with evaluations for each one. The results will be ranked showing you "the winner". If you are satisfied with the result, then the wizard can add generated code to the existing solution if the user wants.

Model Builder is the first step to locate a model/trainer that is suitable for the job. Throughout the book, you shall learn how to use Model Builder to your advantage.

Note   Besides helping locate the best algorithm/model for the current dataset, ML.NET generates very clean code, so that it almost feels like that some expert has written the code.

Summary

This is a short chapter, but I hope it gave you some very top-level view of the ML.NET framework and the rationale behind all kinds of things that are there in the framework. These concepts will be even clearer in the upcoming chapters when we actually create these by hand. In the book, ML and machine learning can be interchangeably used. However, when ML.NET is mentioned, it is specially mentioned with the .NET extension.

CHAPTER 3

Handling Data

Cleansing is engineering

Introduction

Data that are generally available in the real world are not ready for consumption for machine learning activities. A crude but real-life analogy is depicted by the following picture.

Figure 3-1. *Showing data transformation analogy*

However hilarious or not you find this analogy, this is true. The data that are available in the wild need repetitive modifications before it can be fed to a machine learning algorithm; otherwise, the algorithm's performance will take a serious hit and probably be unusable.

The topic of this chapter is to make you acquainted with common practices to deal with different kinds of data to transform those into something that can be given as input to a machine learning algorithm. Some of these techniques are nicely packaged inside the ML.NET library, so you can use it without implementing it yourself.

© Sudipta Mukherjee 2021

S. Mukherjee, *ML.NET Revealed*, https://doi.org/10.1007/978-1-4842-6543-7_3

Before we go deep diving into data transformation, we must know how many types (broad categories) of data are available.

Objective of this chapter

After reading this chapter, you should be able to understand the need for transforming data before feeding those to a machine learning model. You shall also learn how to use several transformations on different kinds of data that ML.NET offers and which one to pick when.

Types of data

Data comes in all different types. Broadly, those are

- Numerical data

- Textual data

- Categorical data

- Location data

- Date and time data (this is also sometimes represented as time series data)

- And so on like images and videos

Numerical data

As the name suggests, numerical data refers to data that are just numbers. Integers and floats are numbers and are thus numerical data. Age of a person, number of visits to the local supermarket per week, number of times someone refuels their car, the amount spent at the movie theater during weekends, and your income are all examples of numerical data.

Textual data

Names, addresses, phone numbers with country codes, email addresses, review comments, feedback messages, comments on social media sites, reviews on movie review sites, and so on are all examples of textual data.

Categorical data

Categorical data is just an enumeration over a preset list. For people like us who are familiar with programming, categorical data is just an incarnation of the Enums in the real world. Names of the blocks in a city, gender (M/F), and postal codes are all examples of categorical data.

The basic difference between textual and categorical data is that textual data is a free form, while categorical data can take one of the many predefined categories as its value.

Location data

As the name suggests, location data is just that; it is data about someone's or some place's location, either expressed in terms of latitude and longitude or via geocodes. It can also be a set of coordinates.

Date and time data

Data about on which day and what time some event occurred.

Transformation of numerical data

Several transformations are available to be performed on numerical data. The goal of all these transformations is to bring data for a given column or columns of numerical data between 0 and 1, which is ideal for the input to a machine learning algorithm that employs some kind of regression; otherwise, the model can be confused because of different scales of different features, and the predicted results will be wrong, more often than acceptable.

Table 3-1. *Showing different normalizing estimators available in ML.NET*

Transformation	Encapsulated as
Mean Normalization	`Transforms.NormalizeMeanVariance`
Log Mean Normalization	`Transforms.NormalizeLogMeanVariance`
Unit Norm Normalization	`Transforms.NormalizeLpNorm`
Global Contrast Normalization	`Transforms.NormalizeGlobalContrast`
Density Normalization	`Transforms.NormalizeBinning`
Density Normalization	`Transforms.NormalizeSupervisedBinning`
Rescaling (min-max normalization)	`Transforms.NormalizeMinMax`

The strategy used in all these normalizations is to dampen/(subtract) the input by the mean (or any other measure) and then normalize the dampened values by variance of any other value, like the maximum value in case of `NormalizeMinMax`.

All these normalization schemes are essentially an estimator that transforms the input data to transformed data as an `IDataView`.

All of these functions have two overloads: one takes an array of `InputOutputColumnPair`, so that you can pass several column names to run the transformation on at one single call. Otherwise, you can run the normalization on a single column once and then use `Append` method to get to the next possible transformations in your pipeline.

```
Microsoft.ML.Transforms.NormalizingEstimator xyz
  = mlContext.Transforms.NormalizeMinMax(new InputOutputColumnPair[]
  {  new InputOutputColumnPair ("horsepower_norm","horsepower"),
     new InputOutputColumnPair ("cylinders_norm")});
```

Figure 3-2. *Showing how to call a normalizing estimator*

One thing to note that although the name of the type is `InputOutputColumnPair`, the parameters for the constructor take the names of the columns in reverse order. The first string passed will be used as the name of the output column, while the second string represents the name of the input column to be transformed. Input column name is droppable, and if dropped, the name of the output column will be used as the input column.

One way to think about normalization is that it is the same as damping. Damping is a physical process where the magnitude of an oscillation reduces with time when no more force is given from outside. Each normalization technique can be thought of as a multiplication of a damping factor to each of the value that produces a new value.

For example, a very common normalization technique is min-max normalization where each value is dampened by the following factor. The first minimum values are subtracted from each value, and then the result is multiplied by the damping factor 1/(`max-min`).

$$x_{normalized} = \frac{x_i - min}{max - min}$$

Figure 3-3. *Showing min-max normalization equation*

In the equation, x_i is the value and min and max are the minimum and maximum values of the column values.

Transformation of categorical data

Machine learning algorithms prefer numeric inputs, and one way to transform categorical data to numeric input is to encode the categorical data to generate a vector. Here is the list of all categorical transformations available in ML.NET.

Table 3-2. *Showing categorical transformation estimators in ML.NET*

Transformation	Encapsulated as
One-hot encoding	mlContext.Transforms.Categorical.OneHotEncoding
One-hot hash encoding	mlContext.Transforms.Categorical.OneHotHashEncoding

One-hot encoding

Encoding categorical variables is a bit tricky. It is tempting to transform a categorical value to a numeric value because it is assumed that machine learning models deal with numeric values. However, this technique will add a bias to the model, and thus resultant predictions will be wrong. Let's say we have a dataset like this.

Table 3-3. *Showing a sample dataset*

CategoricalVal1	Numerical1
A	1.344
B	3.45
M	0.134

If we use Label Encoding and assign one numerical label for each category in Categorical column, then the dataset will look like this.

Table 3-4. *Showing the same dataset with Label Encoding*

CategoricalVal1	Numerical1
1	1.344
2	3.45
3	0.134

But the problem with this is that suddenly for no apparent reason, category "M" will be recognized as a better category than category "A" or "B". If the model performs average internally, then the average of 1 and 3 (representation of "A" and "M") will be category "B", which doesn't make any sense whatsoever. Therefore, we need to transform this dataset to change the rows to columns like this.

Table 3-5. *Table showing result of one-hot encoding applied to sample dataset*

Is_A	Is_B	Is_M	Numerical1
1	0	0	1.344
0	1	0	3.45
0	0	1	0.134

As you can see that now for each category, we have a column that represents the presence of that column in the input dataset. Value of 1 in that column means the presence of that category in the data and value of 0 indicates the absence. So, the first row in the newly created dataset indicates that it is representing category "A" (the hot category, because for this category we have a 1). For the second row, we have B as the hot category. Since in this newly created dataset each row will have exactly one category as set and all others not set, it is known as "one-hot encoding."

One-hot hash encoding

This is the same as the hot encoding, but before the categories are hot encoded, they are hashed using a hash function – thus, the name. Sometimes, there can be multiple incarnations of the same data in categorical data, and using one-hot encoding directly will create more columns in the resultant data than needed and will further confuse the system instead of helping it. In such situations, it is generally a great idea to use a hash function to produce the same hash code for all different looking yet same categorical value. One example of such situation is when we have surnames with slightly different spellings.

Transformation of textual data

Textual data is different than categorical data although it might look similar. Textual data is the free-form text captured as value of a column, while categorical data is the string representation of an enumeration.

Here is the list of all textual data transformations available in ML.NET.

Table 3-6. *Showing different text transformation estimators available in ML.NET*

Transformation	Encapsulated as
FeaturizeText	Transforms.Text.FeaturizeText
TokenizeIntoWords	Transforms.Text.TokenizeIntoWords
TokenizeIntoCharacterAsKeys	Transforms.Text.TokenizeIntoCharacterAsKeys
NormalizeText	Transforms.Text.NormalizeText
ProduceNgrams	Transforms.Text.ProduceNgrams
ProduceWordBags	Transforms.Text.ProduceWordBags
ProduceHashedNgrams	Transforms.Text.ProduceHashedNgrams
RemoveDefaultStopWords	Transforms.Text.RemoveDefaultStopWords
RemoveStopWords	Transforms.Text.RemoveStopWords
LatentDirichletAllocation	Transforms.Text.LatentDirichletAllocation
ApplyWordEmbedding	Transforms.Text.ApplyWordEmbedding

Here are brief details about some of these transformations.

FeaturizeText: This estimator transforms the given input text to a vector of floating-point numbers representing the text. This takes a column name and emits a list of floating-point numbers representing the feature depicted by that column.

NormalizeText: Normalizing the text can among many things mean changing the case of the text, removing punctuations and numbers, and so on. Normalizing texts is required for reconciliation. One example of a sample normalization performed on text is shown as follows:

"Samuel2345." and "Samuel1123;" normalized to remove the numbers and punctuation and lowercased would be "samuel". Reconciliation of addresses and names is quite a challenge and this estimator will be helpful there.

Here is another example of normalizing; this time we remove the space and all punctuations and change the case to uppercase.

"abc def 1234" will become "ABCDEF1234"; so will "abd cef 1234"

The mental map about data handling, cleansing, and augmenting

Figure 3-4. *Data scrubbing*

As the saying goes a picture is worth 1000 words, it is easy to get lost in the literature of these many different functions let alone remember. The following sections attempt to give you some visual clues with some pictures that will possibly strike a chord and remain with you longer than just plain word explanations. Good analogies are really hard to come by; but they are proved to be immensely helpful when trying to grasp difficult/new concepts. Here are some analogies about different techniques to handle data.

In a nutshell, handling data falls into four major different categories.

Table 3-7. *Showing different broad categories of handling data*

Category of operation	Purpose
Normalization	To make every data point in the same range for regressive algorithms
Removal	To remove bad data points
Featurization	To create numerical representation of the data
Missing Value Handling	To augment missing values

Normalization

Figure 3-5. *Showing a pictorial mind-map image for normalization*

Normalizing is almost akin to wood chopping. The proverb size doesn't matter is not appropriate when it comes to feature magnitudes. Analogically speaking, you can imagine the features of the problem domain that you want to feed to your machine learning algorithm, as the wooden logs to be used in a fence. If some of your wooden logs are way too big or way too small than the rest, the fence wouldn't hold off nicely or probably be impossible to build in the first place. Similarly, in a machine learning setting if the scales/magnitudes of the features are way off than others, then the regression-based machine learning models can be really confused and can lead to false/wrong decisions.

Normalizing rightly done always brings down the scale of a feature between 0 and 1. And generally normalization is applicable to numerical data which has a magnitude. Here is the sample of applying min-max normalization before and after:

```
Before (Input) => 1000,2000,1350,2400,1840,1230
After    (Output) => 0, 0.4166667, 0.1458333, 0.5833333, 0.35, 0.09583333
```

As you can see, all the values in the input are normalized to be between 0 and 1.

Removing

Figure 3-6. *Showing a pictorial mind-map image for removal of bad data*

Removing data is like lawn mowing. You just have to get rid of useless weeds (if you will) from your data. Sometimes things to remove could be bad words (a.k.a. stop words) from the textual data. Sometimes it can be removing punctuations or special characters or numbers from textual data. Removing unwanted data (a.k.a. noise) from the data leaves the data in the form that is good for machine learning algorithms to consume.

Sometimes removing is a preparatory step performed on textual data before they can be transformed to numerical representations, and later those numerical representations can be normalized. ML.NET as you have seen in the chapter provides quite some features to remove such things from the data (textual data mostly).

Nowhere other than the search engine, the usage of stop word removal (as a particular form of removal technique) could be seen so effectively. To prove the point that stop words (words that appear way more frequently in every context in a human language than other words, making them least relevant as per information theory) are not important in search, I have searched using these two phrases "Capital of India" and "Capital India". The word "of" is a stop word in English language as it appears almost everywhere without any regard to the context as it is a glue word. So the idea is to show that machine learning algorithms won't be affected if these stop words are dropped; they could be otherwise confused if those words are left as is.

Figure 3-7 shows the search results of *"Capital **of** India"* and *"Capital India"*.

Figure 3-7. *Showing results side by side to show that stop words don't affect search results*

You can see that stop words won't have any effect on the result of the machine learning algorithm. Therefore, those can be safely dropped.

Featurization

Figure 3-8. Showing a sample word cloud that symbolizes featurization

Featurization is like giving a numeric value to the raw data. Unless we could do that, it is impossible to use any machine learning algorithm as you know, and techniques for extracting features from the raw data are a whole new discipline in its own right. *Word cloud* is a very good way to create a *memory map* for what featurization *really is*. As in a word cloud, we give a number (most often the frequency of occurrence) to each word; similarly, we can create numerical representation for all kind of data and that is all featurization is about.

As you have seen, the FeaturizeText method returns a vector of floating-point numbers that represents a text. This is a good example of what featurization can do to data that seemingly have no features except a raw content (for the string in textual data) and the length.

The benefits of featurization are manyfold. The first one is that it really helps in comparison with other seemingly similar things in the world. Comparing two strings for proximity match character by character is a much more computationally involved endeavor than understanding the proximity of two vectors representing those two tokenized strings in N dimension via *cosine similarity* or any other similarity measure. This also has another benefit in that this way, the algorithm becomes scalable.

Handling missing values

Figure 3-9. *Representative image of missing value in a dataset*

Missing values are the real challenge when trying to clean the data to be usable in a machine learning algorithm, because missing values are hard to fill justifiably and with them left as is the performance of the model can be really bad.

There are several strategies to handle missing data. They can be broadly classified into two different categories:

1. Augmenting missing data

2. Removing rows/columns with missing data

Augmenting missing data is a difficult task, primarily because it is difficult to assume what could possibly be a good substitute. There are ways to lamely substitute it with the Maximum, Minimum, or the Default of the column type. ML.NET also provides these in terms of enumerations in ReplacementMode.

```
Microsoft.ML.Transforms
            .MissingValueReplacingEstimator
            .ReplacementMode
```

Figure 3-10. *Showing estimator for handling missing data*

If the column with missing values is numeric, then we can use mean, median, or mode or even an extreme (maximum or minimum) value. However, the decision to replace it with a value will affect the performance, and since one advice is not good for all situation, it is required to do a trial and error analysis to see which missing value replacement strategy is working the best.

For missing values in categorical columns, it is a good idea to mark it with a special value and then use OneHotEncoder.

Handy guide to pick the right transformer/estimator

As the saying goes that if all you have is a hammer, every problem would seem like a nail.

Locating the right function to use can be quite challenging to do the right thing with your data. Table 3-8 tries to ease that a bit.

Table 3-8. *A small cheat sheet to locate the right method for handling your data*

Category	What example best describes your situation	Which function to use
Normalizing	I have features with huge scale differences. Here is an example: one column is number of bedrooms: 1–10, and another is house price "$100000–10000000"	Use any of the normalizing techniques like NormalizeMinMax
Removing	I have textual data with lot of glue words like "if", "of", "for", etc.	RemoveStopWords RemoveDefaultStopWords
Featurize	I have a bunch of movie reviews and I want to make sure how close each one is with the other in terms of their sentiment	FeaturizeText
Normalize	I have a bunch of categorical data. How do I transform them to numeric one?	Don't. Use OneHotEncoding
	I have a bunch of words and I want to extract Ngrams from those words. Ngram is nothing but a list of substring produced by a moving window of a given size over a given string. So for the string "ABCD" and moving window size of 2, Ngram will produce ["AB","BC","CD"]	TokenizeIntoWords
	I have a bunch of Ngrams and I want to featurize those Ngrams	

(*continued*)

Table 3-8. (*continued*)

Category	What example best describes your situation	Which function to use
	I have few different addresses for the same person. All of these addresses are almost the same, and I need to reconcile them to be one	
Missing Value	I need to mark missing values	IndicateMissingValues
	I need to fill in missing values with the minimum/ maximum/default values	MissingValueReplacingEstimator
	I need to fill in missing values with custom values	CustomMapping

Summary

In this chapter, you learned about several data handling and cleansing techniques that ML.NET offers. However, the discussion is kept short here only for the most common type of data encountered in machine learning tasks, namely, numerical and categorical data types. ML.NET offers several other data transformation tasks for image processing, deep learning, and for time series data type. Those are deliberately kept out of this chapter, but I hope that this chapter gives you an essence of what ML.NET provides in terms of data transformations and cleansing and also how these are all glued together using the same estimator, estimator chain, and fitting methods discussed earlier in the book.

ML.NET also offers capabilities to join and drop columns and other related features, but since those were being used throughout the book thus far and will continue to appear in the next chapters, it was not shown again here to save you from boredom.

CHAPTER 4

Regressions

How much will our dream home cost?

Introduction

Ever wondered how can we predict the gasoline price in upcoming months? How the projected exchange rates of currencies are determined? The crux of these problems is the ability to predict a value in a continuous range. The algorithms that solve those problems are called *regression algorithms*. The name regression suggests that these algorithms are mostly iterative in nature. This is different than classification because in classification we need to predict either one of the two values (in case of binary classification) or one of the many (a set of finite labels, in case of multiclass classification) labels. On the other hand, in these situations the predicted value will have to be real value and that's regression. In this chapter, you shall learn about several types of regression algorithms that ML.NET provides and how to measure performance of these algorithms. In some literature, these algorithms are termed as curve fitting algorithms.

Objective

By the end of this chapter, you should be able to identify which problems belong to regression type of problems and solve them using one of the many regression trainers provided by ML.NET. You will also be able to evaluate how good did the algorithm do based on several performance monitoring measures.

Note For the purpose of this book, the terms "Trainers" and "algorithms" are used interchangeably throughout this ML.NET context.

© Sudipta Mukherjee 2021

S. Mukherjee, *ML.NET Revealed*, https://doi.org/10.1007/978-1-4842-6543-7_4

What regression does?

Simple regression is the process of fitting several points in a line. When the line we came up with goes very close to most of the points, this solution is considered good. If the line "misses" many points, we say that the solution is not good. However, there are downsides of both. When almost all the points from the input make their way on the predicted extrapolated (the line that doesn't exist in the input data but projected) line, we guess that we have probably given way too much clue to the model, and thus the model becomes an oracle to give away exact answers for all points from the input dataset. This situation is called *overfitting*, and the reverse when almost no point makes it to the predicted extrapolated line, we call it *underfitting*.

Regression algorithm tries to predict the value of a parameter for a dataset looking at other values. Unlike classification where the label can take either of the two values (binary classification) or many values (multiclass classification), in regression the predicted value is always a real value.

In the simplest case of regression problem, we will have a known parameter (e.g., a number of bedrooms for the house) and a value we want to predict (the price for this house). To do so, the algorithm will have to come up with a formula – an equation, where by setting the known parameter (bedrooms), you will get the value (price). It's logical to expect that the more bedroom a house has, the higher goes the price. So the formula for this relationship can be as follows.

$$y = ax + b$$

Figure 4-1. *Showing equation for the simplest regression*

In real life, we usually have more than one parameter that influences the value we are looking for. For the house price, other parameters could be a safety index of the neighbourhood, total carpet area, age of the property etc. In this case, the equation will look like this.

$$y = a_0 x_0 + a_1 x_1 + ... + a_n x_n + b$$

Figure 4-2. *Generic linear regression*

So the task of a regression algorithm/model is to emit the coefficients or the weights of the input variables. This is still *linear regression* as the nature of the curve is a straight line. However, sometimes a linear model is not enough, and then the following generic equation depicts the nonlinear regression models, also known as *polynomial regression*.

$$y = a_0 x_0 + a_1 x_1^2 + a_2 x_2^2 ... + a_n x_n^n + b$$

Figure 4-3. *Polynomial form*

where b is also known as the regularization term.

Predicting MPG (miles per gallon) for cars

When buying a car, one of the parameters people usually take into consideration is the MPG (miles per gallon of fuel) value. A higher MPG means all other things remaining similar a vehicle is more worth than others. In this experiment, we shall see how we can use `Model Builder` wizard to find the best regression algorithm to predict the MPG value from a dataset.

Figure 4-4. *Representative image of fueling a car*

In this experiment, you shall see how regression can be used to predict the MPG of a used car. You can get the data from `www.kaggle.com/uciml/autompg-dataset`.

The data looks like Figure 4-5.

```
mpg,cylinders,displacement,horsepower,weight,acceleration,model year,origin,car name
18,8,307,130,3504,12,70,1,chevrolet chevelle malibu
15,8,350,165,3693,11.5,70,1,buick skylark 320
18,8,318,150,3436,11,70,1,plymouth satellite
16,8,304,150,3433,12,70,1,amc rebel sst
17,8,302,140,3449,10.5,70,1,ford torino
15,8,429,198,4341,10,70,1,ford galaxie 500
14,8,454,220,4354,9,70,1,chevrolet impala
```

Figure 4-5. *Showing few lines from the mpg dataset*

We shall use Model Builder wizard to get to a decently working model. Follow these steps to get it:

Step 1: In Visual Studio, add machine learning to an existing console app project (Figure 4-6).

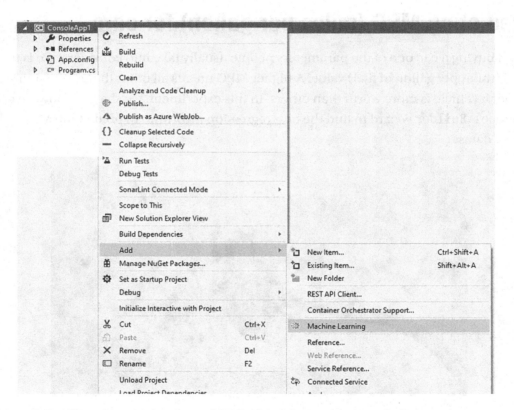

Figure 4-6. *Showing prompt to add Machine Learning to existing project*

Step 2: Select the scenario (Value prediction) for the regression (Figure 4-7).

Select a scenario

Train with your data

The following scenarios use Automated ML to train and pick the best model for your data.
Learn more about training with your own data in Model Builder.

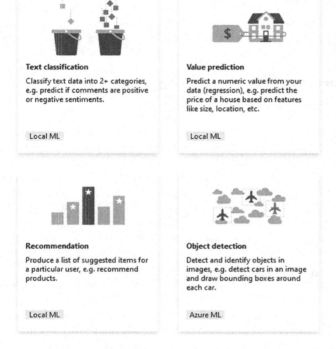

Text classification

Classify text data into 2+ categories, e.g. predict if comments are positive or negative sentiments.

Local ML

Value prediction

Predict a numeric value from your data (regression), e.g. predict the price of a house based on features like size, location, etc.

Local ML

Image classification

Classify images into 2+ categories, e.g. predict whether an image is of a dog or a cat.

Azure ML   Local ML

Recommendation

Produce a list of suggested items for a particular user, e.g. recommend products.

Local ML

Object detection

Detect and identify objects in images, e.g. detect cars in an image and draw bounding boxes around each car.

Azure ML

Figure 4-7. *Select "Value prediction"*

Step 3: Select the file to train the model (Figure 4-8).

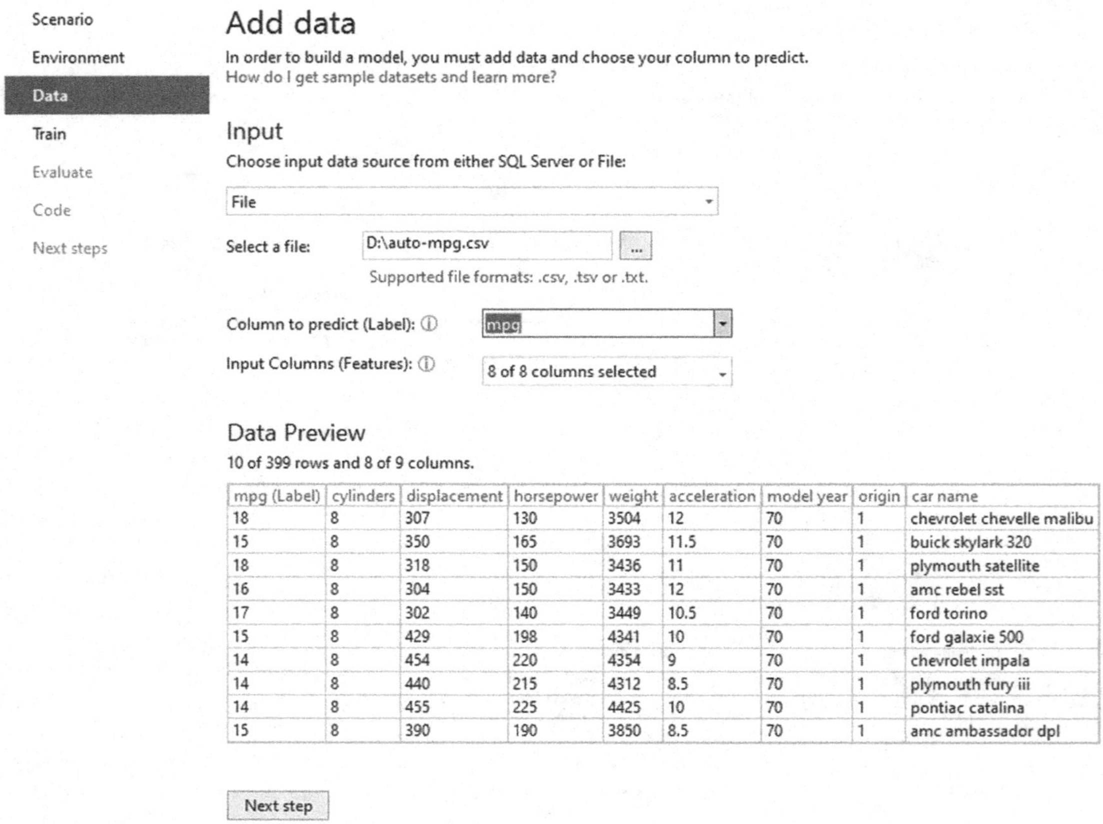

Scenario

Environment

Data

Train

Evaluate

Code

Next steps

Add data

In order to build a model, you must add data and choose your column to predict.
How do I get sample datasets and learn more?

Input

Choose input data source from either SQL Server or File:

| File | ▾ |

Select a file: | D:\auto-mpg.csv | ... |

Supported file formats: .csv, .tsv or .txt.

Column to predict (Label): ⓘ | mpg | ▾ |

Input Columns (Features): ⓘ | 8 of 8 columns selected | ▾ |

Data Preview

10 of 399 rows and 8 of 9 columns.

mpg (Label)	cylinders	displacement	horsepower	weight	acceleration	model year	origin	car name
18	8	307	130	3504	12	70	1	chevrolet chevelle malibu
15	8	350	165	3693	11.5	70	1	buick skylark 320
18	8	318	150	3436	11	70	1	plymouth satellite
16	8	304	150	3433	12	70	1	amc rebel sst
17	8	302	140	3449	10.5	70	1	ford torino
15	8	429	198	4341	10	70	1	ford galaxie 500
14	8	454	220	4354	9	70	1	chevrolet impala
14	8	440	215	4312	8.5	70	1	plymouth fury iii
14	8	455	225	4425	10	70	1	pontiac catalina
15	8	390	190	3850	8.5	70	1	amc ambassador dpl

Next step

Figure 4-8. *Showing the training data loaded*

Step 4: Start the training. Leaving the training phase in Model Builder for longer really gives better results (I recommend 2 minutes at least). The official documentation says the following.

0 - 10 MB ➔ 10 sec

10 - 100 MB ➔   10 min

100 - 500 MB ➔ 30 min

500 - 1 GB ➔ 60 min

Figure 4-9. *Recommended time required for training by Microsoft*

Even though the suggested time according to the documentation is 10 seconds, I would recommend to run the training for longer around 2 minutes (this is found from my personal experience, could vary depending on your PC hardware) because that gives the time to find the best possible algorithm.

Step 5: As the program (Model Builder wizard) runs, it will show the status (progress) of the execution (Figure 4-10).

Figure 4-10. *Final result of the training*

At the end, the program reports the final performance and the best algorithm/model for this dataset.

Step 6: Check out the evaluation report provided by the wizard (Figure 4-11).

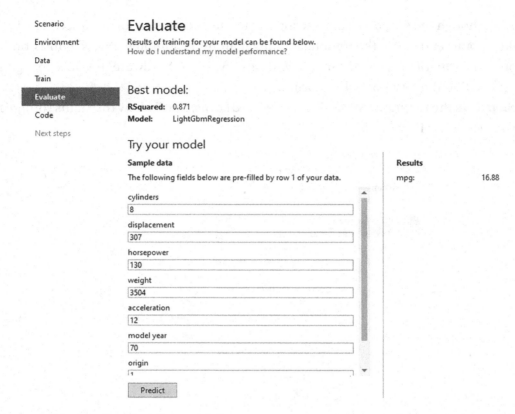

Figure 4-11. *Evaluating the model*

The evaluation table here shows several performance measurement metrics for the model.

Step 7: The generated code will be automatically added to the host solution.

Step 8: The current version of ML.NET offers a way to read more and deploy the model as ASP.NET service.

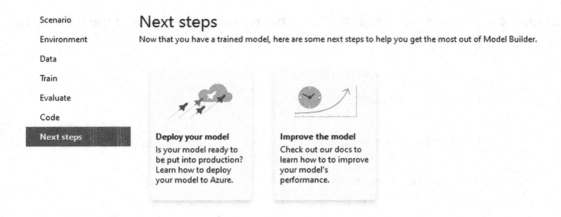

Scenario

Environment

Data

Train

Evaluate

Code

Next steps

Next steps

Now that you have a trained model, here are some next steps to help you get the most out of Model Builder.

Deploy your model

Is your model ready to be put into production? Learn how to deploy your model to Azure.

Improve the model

Check out our docs to learn how to to improve your model's performance.

***Figure 4-12.** Next steps wizard*

Step 9: Look at the generated code.

If you choose to add the generated projects, the wizard will add a couple of projects to the solution as shown in Figure 4-13.

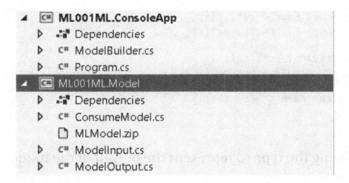

***Figure 4-13.** Autogenerated code added to the existing solution*

The ML.Model project holds types to represent one row of the input data (ModelInput.cs) and the prediction ModelOutput.cs. Here are these two generated types (Listings 4-1 and 4-2).

Listing 4-1. Generated `ModelInput.cs` that represents one row of the input data

```
public class ModelInput
{
    [ColumnName("mpg"), LoadColumn(0)]
    1 reference
    public float Mpg { get; set; }
    [ColumnName("cylinders"), LoadColumn(1)]
    1 reference
    public float Cylinders { get; set; }
    [ColumnName("displacement"), LoadColumn(2)]
    1 reference
    public float Displacement { get; set; }
    [ColumnName("horsepower"), LoadColumn(3)]
    1 reference
    public float Horsepower { get; set; }
    [ColumnName("weight"), LoadColumn(4)]
    1 reference
    public float Weight { get; set; }
    [ColumnName("acceleration"), LoadColumn(5)]
    1 reference
    public float Acceleration { get; set; }
    [ColumnName("model year"), LoadColumn(6)]
    1 reference
    public float Model_year { get; set; }
    [ColumnName("origin"), LoadColumn(7)]
    1 reference
    public float Origin { get; set; }
    [ColumnName("car name"), LoadColumn(8)]
    1 reference
    public string Car_name { get; set; }
}
```

Listing 4-2. Showing the type to represent the output of the model

```
4 references
public class ModelOutput
{
    1 reference
    public float Score { get; set; }
}
```

Notice that since the output of a regression model is real value, the property Score represents that.

Code walkthrough

The magic happens in the `BuildPipeline` method. Here is the code after a bit of formatting to make it more readable (Listing 4-3).

Listing 4-3. Showing how the pipeline is being built

```
1 reference
public static IEstimator<ITransformer> BuildTrainingPipeline(MLContext mlContext)
{
    // Data process configuration with pipeline data transformations
    var dataProcessPipeline = mlContext.Transforms.IndicateMissingValues(new[]
            { new InputOutputColumnPair("horsepower_MissingIndicator", "horsepower") })
        .Append(mlContext.Transforms.Conversion.ConvertType(
            new[] { new InputOutputColumnPair("horsepower_MissingIndicator",
                                              "horsepower_MissingIndicator") }))
        .Append(mlContext.Transforms.ReplaceMissingValues(
            new[] { new InputOutputColumnPair("horsepower", "horsepower") }))
        .Append(mlContext.Transforms.Concatenate("Features",
            new[] { "horsepower_MissingIndicator",
                    "horsepower",
                    "cylinders", "displacement", "weight", "acceleration", "model year", "origin" }));

    // Set the training algorithm
    var trainer = mlContext.Regression.Trainers.FastTreeTweedie(new FastTreeTweedieTrainer.Options()
                    {
                        NumberOfLeaves = 7,
                        MinimumExampleCountPerLeaf = 1,
                        NumberOfTrees = 100,
                        LearningRate = 0.1908787f,
                        Shrinkage = 0.3315645f,
                        LabelColumnName = "mpg",
                        FeatureColumnName = "Features"
                    });

    var trainingPipeline = dataProcessPipeline.Append(trainer);
```

ML.NET model builder wizard really does quite an impressive job. It not only creates the model but also generates code that is highly readable and maintainable by future programmers who would otherwise be scratching their heads. I must admit for generated code, ML.NET generated code looks really nice.

The "horsepower" column had some missing values, and ML.NET figures that out and applies a couple of transformations:

> `IndicateMissingValues`: To mark the column "horsepower" to have missing values

> `ReplaceMissingValues`: To replace the missing values with a predefined value

63

At the end of the pipeline, `FastTreeTweedie` trainer is used, which performs well if there are many zeros or missing values.

Predicting house prices in Boston suburbs

In the following experiment, you shall learn about the steps to use ML.NET to predict house prices in Boston suburbs. The example might be a toyish one, but the learning is transferable to a more production-ready environment.

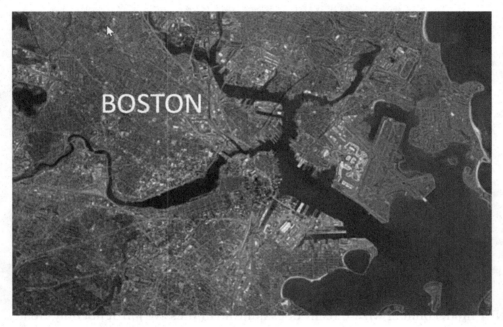

Figure 4-14. *Representative aerial view of Boston*

The Boston Housing Dataset is a derived from information collected by the US Census Service concerning housing in the area of Boston MA. The following describes the dataset columns (this list is taken from the description on Kaggle):

- **CRIM**: Per capita crime rate by town

- **ZN**: Proportion of residential land zoned for lots over 25,000 sq. ft.

- **INDUS**: Proportion of nonretail business acres per town

- **CHAS**: Charles River dummy variable (1 if tract bounds river; 0 otherwise)

- **NOX**: Nitric oxides concentration (parts per 10 million)

- **RM**: Average number of rooms per dwelling

- **AGE**: Proportion of owner-occupied units built prior to 1940

- **DIS**: Weighted distances to five Boston employment centers

- **RAD**: Index of accessibility to radial highways

- **TAX**: Full-value property-tax rate per $10,000

- **PTRATIO**: Pupil-teacher ratio by town

- **B**: 1000(Bk - 0.63)^2 where Bk is the proportion of blacks by town

- **LSTAT**: % lower status of the population

- **MEDV** Median value of owner-occupied homes in $1000's

And this time we shall not use Model Builder wizard but hand code our model, changing one trainer at a time.

Step 1: Create a new console project.

Step 2: Create the following class `BostonHouse.cs` (this is the Model input).

Listing 4-4. ModelInput for Boston housing problem

```
public class BostonHouse
 {
     /// <summary>
     /// CRIM - per capita crime rate by town
     /// </summary>
     [LoadColumn(0), ColumnName("crim") ]
     public float CRIM { get; set; }
     /// <summary>
     /// proportion of residential land zoned for lots over
         25,000 sq. ft.
     /// </summary>
     [LoadColumn(1), ColumnName("zn")]
     public float ZN { get; set; }
```

```
/// <summary>
/// proportion of nonretail business acres per town
/// </summary>
[LoadColumn(2), ColumnName("indus")]
public float INDUS { get; set; }
/// <summary>
/// Charles River dummy variable (1 if tract bounds river; 0
otherwise)
/// </summary>
[LoadColumn(3),ColumnName("chas")]
public float CHAS { get; set; }
/// <summary>
/// nitric oxides concentration (parts per 10 million)
/// </summary>
[LoadColumn(4), ColumnName("nox")]
public float NOX { get; set; }
/// <summary>
/// average number of rooms per dwelling
/// </summary>
[LoadColumn(5), ColumnName("rm")]
public float RM { get; set; }
/// <summary>
/// proportion of owner-occupied units built prior to 1940
/// </summary>
[LoadColumn(6), ColumnName("age")]
public float Age { get; set; }
/// <summary>
/// weighted distances to five Boston employment centers
/// </summary>
[LoadColumn(7), ColumnName("dis")]
public float DIS { get; set; }
/// <summary>
/// index of accessibility to radial highways
/// </summary>
[LoadColumn(8),ColumnName("rad")]
```

```
    public float RAD { get; set; }
    /// <summary>
    /// full-value property-tax rate per $10,000
    /// </summary>
    [LoadColumn(9), ColumnName("tax")]
    public float TAX { get; set; }
    /// <summary>
    /// pupil-teacher ratio by town
    /// </summary>
    [LoadColumn(10) , ColumnName("ptratio")]
    public float PTRATIO { get; set; }
    /// <summary>
    /// 1000(Bk - 0.63)^2 where Bk is the proportion of blacks by town
    /// </summary>
    [LoadColumn(11), ColumnName("b")]
    public float B { get; set; }
    /// <summary>
    /// % lower status of the population
    /// </summary>
    [LoadColumn(12), ColumnName("lstat")]
    public float LSTAT { get; set; }

    [LoadColumn(13), ColumnName("medv")]
    public float Medv { get; set; }
}
```

Step 3: Create the following class BostonHousePrice.cs (this is the Model output).

Listing 4-5. ModelOutput for Boston housing problem

```
public class BostonHousePrice
{
    public float MEDV { get; set; }
}
```

Step 4: Add the following lines in the Program.cs.

67

Listing 4-6. Consuming the Boston housing price prediction regression model

```
//change your path accordingly
string DATA_FILEPATH = @"C:\MLDOTNET\housing.csv";
MLContext context = new MLContext(seed: 1);

IDataView trainingDataView = context.Data.LoadFromTextFile<BostonHouse>(
                                   path: DATA_FILEPATH,
                                   hasHeader: true,
                                   separatorChar: ',',
                                   allowQuoting: true,
                                   allowSparse: false);

var pipeLine = context.Transforms.NormalizeMinMax("crim", "crim")
              .Append(context.Transforms.NormalizeMinMax("zn", "zn"))
              .Append(context.Transforms.NormalizeMinMax("indus",
               "indus"))
              .Append(context.Transforms.NormalizeMinMax("chas", "chas"))
              .Append(context.Transforms.NormalizeMinMax("nox", "nox"))
              .Append(context.Transforms.NormalizeMinMax("rm", "rm"))
              .Append(context.Transforms.NormalizeMinMax("age", "age"))
              .Append(context.Transforms.Concatenate("Features",
               "crim", "zn", "indus", "chas", "nox", "rm", "age"));
          // Set the training algorithm
var trainer = context.Regression.Trainers.Sdca(labelColumnName: "medv");
var trainingPipeline = pipeLine.Append(trainer);

var model = trainingPipeline.Fit(trainingDataView);
var engine = context.Model
    .CreatePredictionEngine<BostonHouse, BostonHousePrice>(model);

var input = CreateSingleDataSample(DATA_FILEPATH);
var result = engine.Predict(input);

Console.WriteLine($"Actual MEDV is {sampleData.Medv}");
Console.WriteLine($"Predicted MEDV is {result.Medv}");
```

This produces the following output:

Actual MEDV is 24
Predicted MEDV is 26.32589

This program uses `NormalizeMinMax` transformations on the numeric columns.

Performance metrics

All performance metrics are available in `RegressionMetrics` class of `Microsoft.ML.Data` namespace as shown in Figure 4-15.

```
namespace Microsoft.ML.Data
{
    ...public sealed class RegressionMetrics
    {
        ...public double MeanAbsoluteError { get; }
        ...public double MeanSquaredError { get; }
        ...public double RootMeanSquaredError { get; }
        ...public double LossFunction { get; }
        ...public double RSquared { get; }
    }
}
```

Figure 4-15. *Showing evaluation matrices*

R-squared is a statistical measure that represents the goodness of fit of a regression model. The ideal value for r-squared is 1. The closer the value of R-square to 1, the better is the model fitted. This metric is available as "RSquared" in the `ML.Data`.

R-square is a comparison of residual sum of squares (`SSres`) with total sum of squares (`SStotal`). The total sum of squares is calculated by summation of squares of perpendicular distance between data points and the average line (Figure 4-16).

$$SS_{total} = \sum (y_i - y_{avg})^2$$

Figure 4-16. *Sum of squares equation*

The residual sum of squares is calculated by the summation of squares of perpendicular distance between data points and the best fitted line (Figure 4-17).

$$SS_{res} = \sum (y_i - y^{\wedge})^2$$

Figure 4-17. *Residual sum of total equation*

Joining this together, R-squared is given by the formula shown in Figure 4-18.

$$R^2 = 1 - \frac{SS_{res}}{SS_{tot}}$$

Figure 4-18. *R-squared equation*

If the value of R-squared error approaches 1, then the regression is achieving good result.

Mean squared error

This is the average of the squared differences between actual and predicted values. All the negatives are dampened because of the square. In other words, due to a square negative values become positive which allows to track the accumulative difference. Therefore, the actual amplitude of the error is considered.

$$\frac{1}{n} \sum_{i=1}^{n} (Y_i - \hat{Y}_i)^2$$

$* n$ is the number of data points
$* Y_i$ represents observed values
$* \hat{Y}_i$ represents predicted values

Figure 4-19. *Mean squared error equation*

Root mean square

As the name suggests, it is the root of the mean squared difference of the predicted and actual value. This is further damped or regularized error and generally leaves less room for getting more errors. This error makes big errors not to confuse the model.

$$RMSE = \sqrt{\sum_{i=1}^{n} \frac{(\hat{y}_i - y_i)^2}{n}}$$

Figure 4-20. *Root mean square error equation*

Normalized root mean square

This error metric is really useful for comparison of several models which have features on different scales. This is not readily available from ML.NET, but as you can see from the formula, it can be easily calculated (Figure 4-21).

$$NRMSD = \frac{RMSD}{y_{max} - y_{min}} \text{ or } NRMSD = \frac{RMSD}{\bar{y}}.$$

Figure 4-21. *Equation of NRMSD*

Regression Trainer	Encapsulated as
Fast Tree	Regression.Trainers.FastTree
Fast Forest	Regression.Trainers.FastForest
Fast Tree Tweedie	Regression.Trainers.FastTreeTweedie
Generalized Additive Models	Regression.Trainers.Gam
Limited-Memory BFGSPoissonRegression	Regression.Trainers.LbfsgPoissionRegression
Online Gradient Descent	Regression.Trainers.OnlineGradientDescent
Sdc	Regression.Trainers.Sdca

Ideas of using regression to improve your daily life

Predict time to reach work/school depending on when you leave your home.

Time taken to drive to school!

If data is maintained for several months about the traffic situation, weather, holidays, and time taken to reach school, then a regression model can take this data and predict the possible time it will take to reach school

Recommending grocery to buy

If data is maintained for several months about the grocery ordering pattern of a family then not-too-long after we can start predicting the quantity of food/grocery they would need beforehand using regression techniques.

Summary

In this chapter, you have learned about how to use ML.NET for regressions and how to check the performance of the model arrived. In the next chapter, you shall learn about classification algorithms that ML.NET offers. I hope this chapter is leaving you with enough motivation to try different algorithms to address regression problems in your job/life.

CHAPTER 5

Classifications

Helping computers tell chalk and cheese apart

Introduction

One of the major supervised machine learning class of problems is to classify things from a set of given things, by learning from previously labeled data. This is like proverbial *"Telling chalk and cheese apart"* from several examples of labeled data. In this chapter, we shall go through an example of classification problem and will solve it using ML.NET.

Objective of this chapter

The objective is to give you a good understanding of classification type of problems and introduce several trainers/classifiers available in ML.NET for this type of problems. I'll demonstrate how to solve a classification problem in ML.NET Model Builder and how to configure and use the classification trainers.

By the end of this chapter, you shall be able to view a problem presented as a classification problem and use any of the available classifier to solve it. You shall also be able to evaluate the performance of the classifier and tune it if required before deploying in production environment.

Types of classifications

There are two types of classification problems that can manifest in the wild. When the task is to tag an unknown entry with one of the two possible classes/types from the previously presented labeled data, then that task is called *binary classification*. On the

© Sudipta Mukherjee 2021
S. Mukherjee, *ML.NET Revealed*, https://doi.org/10.1007/978-1-4842-6543-7_5

other hand, if the task is to predict the confidence of the model as to which of the many different types the unknown entry possibly belongs to, it is called *multiclass classification* for obvious reasons.

For example, predicting whether the animal in each image is a dog or a cat is an example of binary classification, while identifying handwritten digits to be one of 0 to 9 is a case of multiclass classification problem. As you might imagine rightly, binary classification is a generalization of multiclass classification problem where there are just two types or classes that an unknown entry can belong to.

Terminologies of data

In all supervised algorithm, we need data to train the system and data to test the performance of the system.

Training data: Data that is used to create the model that will predict the result

Test data: Data that is used to check the performance of the model

Ideally, training and test data should be sourced differently and shouldn't overlap intentionally. However, most of the time data is not available to test the performance of the system, and in these occasions, one can use a part of the training data as test data. This split between training and test is often called train-test-split.

Example case studies

In the following sections, several example situations of case study of classifications are presented. And ML.NET is used to craft a solution.

Using ML.NET for predicting whether income will be more than 50K USD. The task is to predict whether a given individual will be able to earn more than 50K or not based on other demographic features. The data can be downloaded from `https:// archive.ics.uci.edu/ml/machine-learning-databases/adult/`.

Here are the first few rows of the dataset (Figure 5-1).

```
39, State-gov, 77516, Bachelors, 13, Never-married, Adm-clerical, Not-in-family, White, Male, 2174, 0, 40, United-States, <=50K
50, Self-emp-not-inc, 83311, Bachelors, 13, Married-civ-spouse, Exec-managerial, Husband, White, Male, 0, 0, 13, United-States, <=50K
38, Private, 215646, HS-grad, 9, Divorced, Handlers-cleaners, Not-in-family, White, Male, 0, 0, 40, United-States, <=50K
53, Private, 234721, 11th, 7, Married-civ-spouse, Handlers-cleaners, Husband, Black, Male, 0, 0, 40, United-States, <=50K
28, Private, 338409, Bachelors, 13, Married-civ-spouse, Prof-specialty, Wife, Black, Female, 0, 0, 40, Cuba, <=50K
37, Private, 284582, Masters, 14, Married-civ-spouse, Exec-managerial, Wife, White, Female, 0, 0, 40, United-States, <=50K
49, Private, 160187, 9th, 5, Married-spouse-absent, Other-service, Not-in-family, Black, Female, 0, 0, 16, Jamaica, <=50K
52, Self-emp-not-inc, 209642, HS-grad, 9, Married-civ-spouse, Exec-managerial, Husband, White, Male, 0, 0, 45, United-States, >50K
31, Private, 45781, Masters, 14, Never-married, Prof-specialty, Not-in-family, White, Female, 14084, 0, 50, United-States, >50K
```

Figure 5-1. *Showing raw data of salary segregation*

The dataset doesn't come with headers. The headers are available in the adult.names as shown in the highlighted box in Figure 5-2.

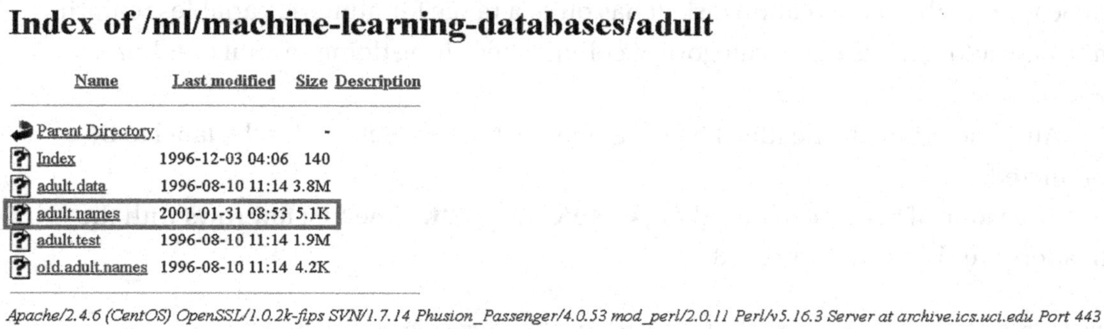

Index of /ml/machine-learning-databases/adult

Name	Last modified	Size	Description
Parent Directory		-	
Index	1996-12-03 04:06	140	
adult.data	1996-08-10 11:14	3.8M	
adult.names	2001-01-31 08:53	5.1K	
adult.test	1996-08-10 11:14	1.9M	
old.adult.names	1996-08-10 11:14	4.2K	

Apache/2.4.6 (CentOS) OpenSSL/1.0.2k-fips SVN/1.7.14 Phusion_Passenger/4.0.53 mod_perl/2.0.11 Perl/v5.16.3 Server at archive.ics.uci.edu Port 443

Figure 5-2. *The download page for the dataset with salary information*

The content of the **adult.names** files has the headers. Here are the headers. For space constraint, I have not shown the values of each column or their types.

- age
- workclass
- fnlwgt
- education
- education-num
- marital-status
- occupation
- relationship
- race
- sex

- capital-gain

- capital-loss

- hours-per-week

- native-country

The dataset is very interesting as an example, because it has everything you want to try out a machine learning algorithm for classification. It has missing data (marked with "?" symbols). It has quite a good mix of numeric and categorical data in the mix to be used in the classification task. It has quite a range for numeric variables requiring normalization. It has many categorical columns requiring doing several one-hot encodings.

After including the header, I have named the header "Salary" for the label to be predicted.

The value of Salary can be either "**<=50K**" or "**>50K**". The first few rows with the headers are shown in Figure 5-3.

```
age,workclass,fnlwgt,education,education-num,marital-status,occupation,relationship,race,sex,capital-gain,capital-loss,hours-per-week,native-country,Salary
39, State-gov, 77516, Bachelors, 13, Never-married, Adm-clerical, Not-in-family, White, Male, 2174, 0, 40, United-States, <=50K
50, Self-emp-not-inc, 83311, Bachelors, 13, Married-civ-spouse, Exec-managerial, Husband, White, Male, 0, 0, 13, United-States, <=50K
38, Private, 215646, HS-grad, 9, Divorced, Handlers-cleaners, Not-in-family, White, Male, 0, 0, 40, United-States, <=50K
53, Private, 234721, 11th, 7, Married-civ-spouse, Handlers-cleaners, Husband, Black, Male, 0, 0, 40, United-States, <=50K
28, Private, 338409, Bachelors, 13, Married-civ-spouse, Prof-specialty, Wife, Black, Female, 0, 0, 40, Cuba, <=50K
37, Private, 284582, Masters, 14, Married-civ-spouse, Exec-managerial, Wife, White, Female, 0, 0, 40, United-States, <=50K
49, Private, 160187, 9th, 5, Married-spouse-absent, Other-service, Not-in-family, Black, Female, 0, 0, 16, Jamaica, <=50K
52, Self-emp-not-inc, 209642, HS-grad, 9, Married-civ-spouse, Exec-managerial, Husband, White, Male, 0, 0, 45, United-States, >50K
31, Private, 45781, Masters, 14, Never-married, Prof-specialty, Not-in-family, White, Female, 14084, 0, 50, United-States, >50K
```

Figure 5-3. *Salary data annotated with their headers*

This dataset along with headers is now ready for ML.NET model builder. The following section shows how to feed this data to model builder to get the initial sketch of the learning system.

Step 1: Add a machine learning to an existing project (Figure 5-4).

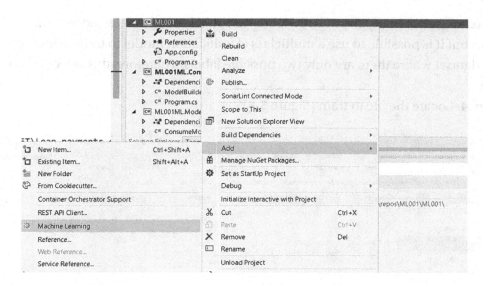

Figure 5-4. *Creating a new Machine Learning project*

Step 2: Select the scenario for which you want to train. In this case, you can select Issue Classification (Figure 5-5).

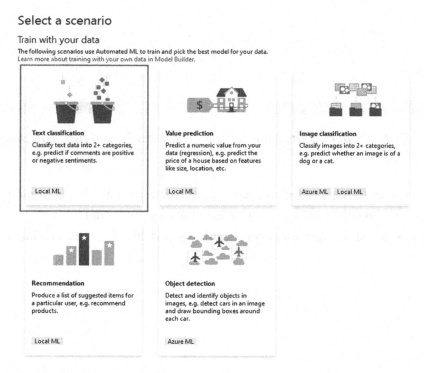

Figure 5-5. *Interface for selecting the type of Machine Learning*

Here, we select Issue Classification, but we could have also selected Sentiment Analysis but it is possible to use a multiclass classification model to train to identify a binary dataset where there are only two possible labels. Therefore, this selection is good for use.

Step 4: Locate the file to train (Figure 5-6).

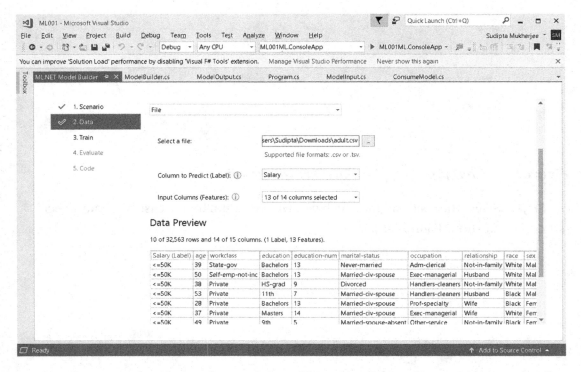

Figure 5-6. *Data loaded in the Model Builder wizard (Configuration pane for the training)*

Once you do so, the Model builder will load the data from the file as shown, and then you can select which columns you want to use to train your model.

Step 4: Let the system train for about 2 minutes and click the "Start training" button (Figure 5-7).

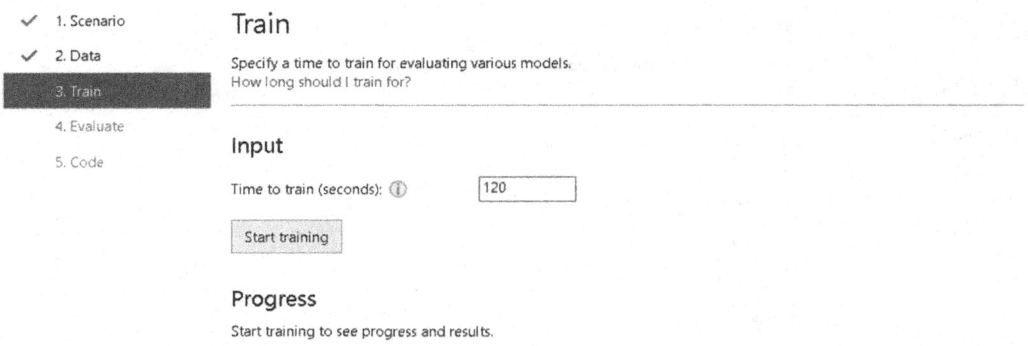

Figure 5-7. *Configuration for training duration in Model Builder wizard*

Step 5: Wait for the system to train and check the progress as shown in Figure 5-8.

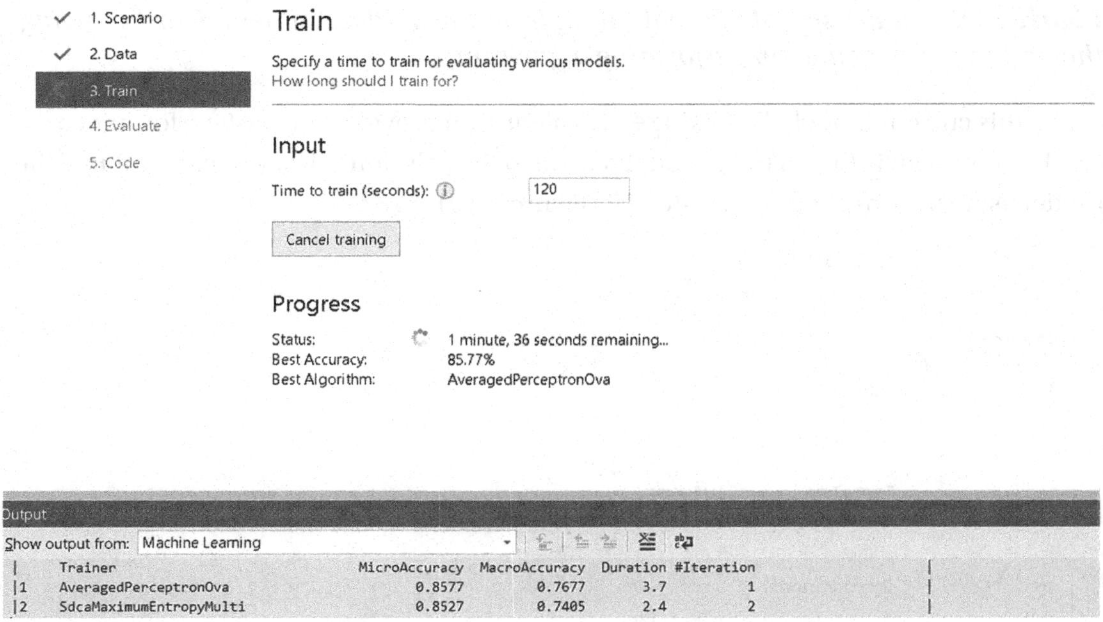

	Trainer	MicroAccuracy	MacroAccuracy	Duration	#Iteration	
1	AveragedPerceptronOva	0.8577	0.7677	3.7	1	
2	SdcaMaximumEntropyMulti	0.8527	0.7405	2.4	2	

Figure 5-8. *Showing progress of the training (still in progress)*

The model builder shows the performance of the classifiers tried on so far.

Once the model builder is successfully completed running, it will show the result of the finalized model as shown in Figure 5-9.

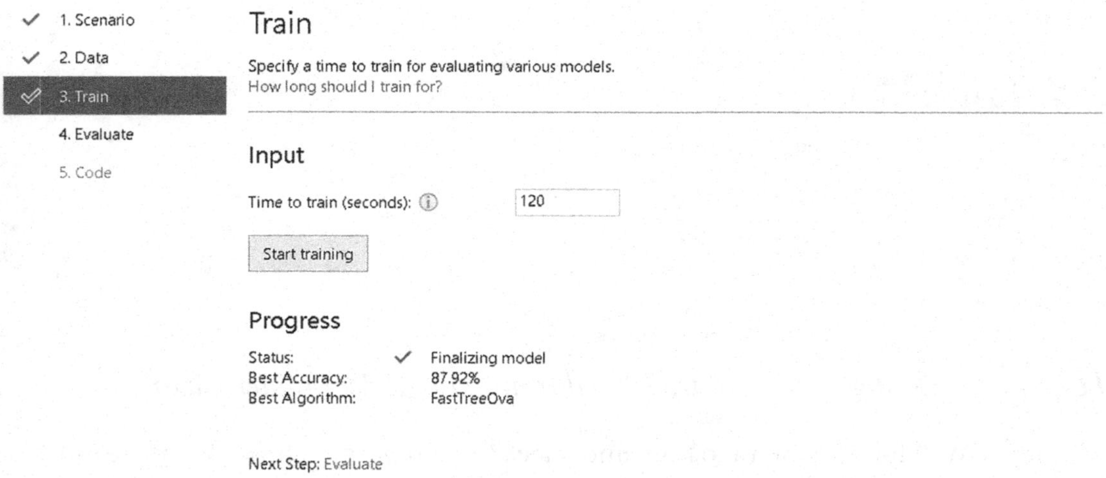

Figure 5-9. *Final result of the evaluation from the Model Builder wizard showing the performance of the top performing algorithm*

In this current model, the `FastTreeOva` algorithm provides the best performance.

The next step is to add the generated code to the solution. Also, you can evaluate the model using the on-the-fly generated UI (Figure 5-10).

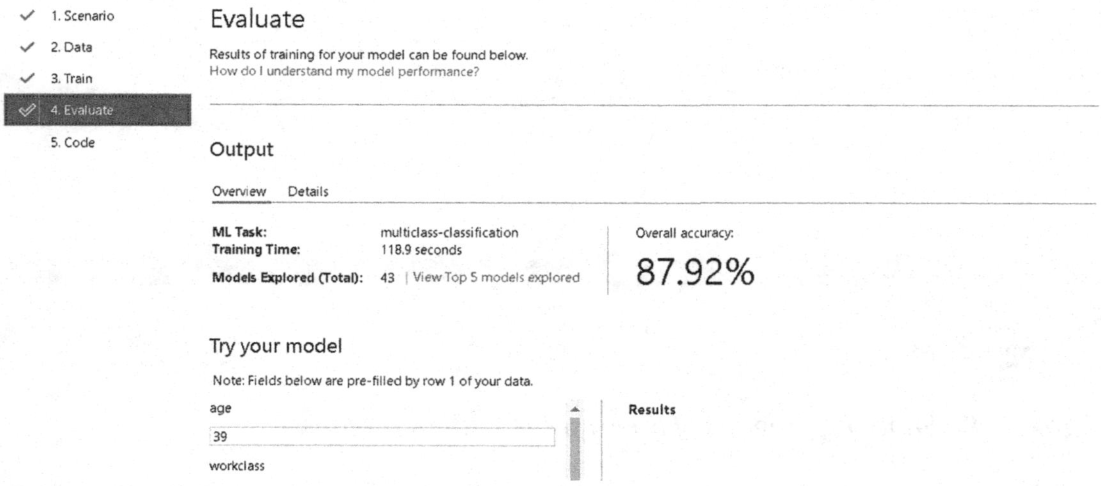

Figure 5-10. *Evaluate tab on Model Builder showing the overall accuracy of the best discovered model*

This is a very nice dashboard showing

- The final score of the model's accuracy

- An interface generated on-the-fly to try out the model

- Total number of models tried (43 in this case; imagine how long it would have taken to try those manually)

The logical next step (if you are mostly satisfied) with the model is to add the generated projects to the solution as indicated by step 5 on the wizard (Figure 5-11).

Figure 5-11. *Model Builder generated code add prompt interface*

Once done, a couple of projects will get added to the solution as explained by the wizard (Figure 5-12).

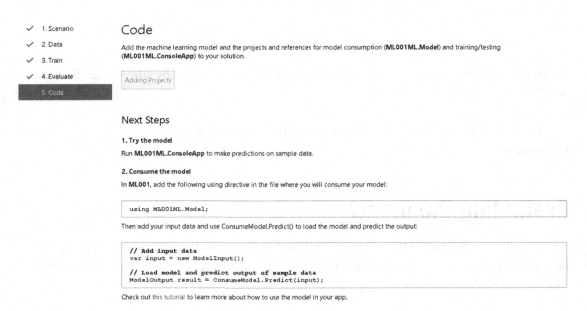

Figure 5-12. *Showing explanation of the generated code to be added*

Here, a couple of projects are shown that get added to your existing solution (Figure 5-13).

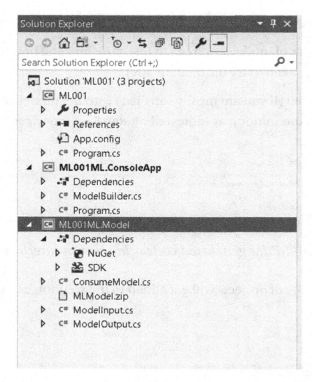

Figure 5-13. *Showing a couple of generated projects added by the Model Builder wizard*

The .ConsoleApp is the client application that shows how to consume the model that is generated.

Evaluating the model

There is also a class generated called ModelBuilder.cs and that has all the logic to see how good or bad the current model performed. There are several matrices to determine that.

Confusion matrix

As the name suggests, the confusion matrix is a measure of how confused or not the algorithm while predicting the labels of the inputs. Confusion matrix consists of basically four values:

- TP (true positives)

- FP (false positives)

- TN (true negatives)

- FN (false negatives)

This picture nicely captures the essence of confusion matrix (Figure 5-14).

		Actual class		
		Cat	Dog	Rabbit
Predicted class	**Cat**	5	2	0
	Dog	3	3	2
	Rabbit	0	1	11

Figure 5-14. *A sample confusion matrix of a classification task that tries to identify cats, dogs, and rabbits*

- ***TP***: The set of predictions where the classification labels an actual true label as true.

- ***TN***: The set of predictions where the classification labels an actual false label as false.

- ***FP***: The set of predictions where the classification labels an actual false label as true. This is known as Type I error.

- ***FN***: The set of predictions where the classification labels an actual true label as false. This is known as Type II error.

Two very important measures are calculated from the confusion matrix. These are called precision and recall. These are calculated as per the following formulae:

$$precision = \frac{TP}{TP + FP}$$

$$recall = \frac{TP}{TP + FN}$$

It is difficult to compare multiple models with obviously different precision and recall. Therefore, their harmonic mean is calculated, and this is a very popular measure to determine the performance of the model. This measure is called the F1 score. The formula is given as follows:

$$F1\_Score = \frac{2 * precision * recall}{precision + recall}$$

The higher the F1 score, the better. To see the confusion matrix for the problem solved, add the following lines in the `PrintMulticlassClassification FoldsAverageMetrics` method:

```
var confusionMatrices = crossValResults.Select(r => r.Metrics.
ConfusionMatrix);

foreach (var confusionMat in confusionMatrices)
{
    Console.WriteLine(confusionMat.GetFormattedConfusionTable());
}
```

This will print all the confusion matrix as shown in Figure 5-15 (only two are shown for space constraint).

```
Confusion table
         ||======================
PREDICTED ||    <=50K |      >50K | Recall
TRUTH    ||======================
   <=50K || TP  4,758 | FN    256 | 0.9489
    >50K || FP    569 | TN    969 | 0.6300
         ||======================
Precision||    0.8932 |    0.7910 |


Confusion table
         ||======================
PREDICTED ||    <=50K |      >50K | Recall
TRUTH    ||======================
   <=50K ||    4,631 |        254 | 0.9480
    >50K ||      578 |      1,035 | 0.6417
         ||======================
Precision||    0.8890 |    0.8029 |
```

Figure 5-15. *Showing formatted confusion matrices for the salary prediction problem solved*

The second confusion matrix shows that it predicted 4631 entries as <= 50 and that is actually true, and 1035 entries are predicted as >50K and which is also actually right. Therefore, TP = 4631 and >50K is identified as <=50K in 578 cases and FP = 578 and FN = 254. Using the formula, recall is TP/(TP+FN) => 0.9480 and precision is TP/(FP+TP) => 0.8890.

I have annotated the boxes in the image with TP and so on in yellow as a memory map. It is easy to remember that the parts of the matrix are given in four quadrants available in a counterclockwise manner starting from top left with TP, FP, TN, and FN. The first two quadrants belong to the positive results, while the last two to the negative results, and it is true cases followed by the false cases in both occasions.

> *Micro Accuracy*: How often we get the right answer from the model. If you want to calculate only one metric for checking the performance of your classification algorithm, then use this metric.

> *Macro Accuracy*: This is basically the average of micro accuracies computed for each class/label in the dataset.

Log loss

Logarithmic loss (related to cross-entropy) measures the performance of a classification model where the prediction input is a probability value between 0 and 1. The goal of our machine learning models is to minimize this value. A perfect model would have a log loss of 0. Log loss increases as the predicted probability diverges from the actual label. So, predicting a probability of .012 when the actual observation label is 1 would be bad and result in a high log loss.

Formulae for log loss:

For binary classification

$$-\big(y * log(p) + (1-y)log(1-p)\big)$$

For multiclass classification

$$-\sum_{c=1}^{M} y_{0,c} \log\big(p_{0,c}\big)$$

To see how the model performed, a call to the CreateModel method from the ML.ConsoleApp application can be made as this

```
ModelBuilder.CreateModel();
```

And this will generate results like this for you (Figure 5-16).

```
=============== Cross-validating to get model's accuracy metrics ===============
********************************************************************************
*       Metrics for Multi-class Classification model
*------------------------------------------------------------------------------
*       Average MicroAccuracy:      0.87  - Standard deviation: (.003)  - Confidence Interval 95%: (.003)
*       Average MacroAccuracy:      0.788 - Standard deviation: (.007)  - Confidence Interval 95%: (.007)
*       Average LogLoss:            .296  - Standard deviation: (.004)  - Confidence Interval 95%: (.003)
*       Average LogLossReduction:   .463  - Standard deviation: (.009)  - Confidence Interval 95%: (.009)
********************************************************************************
```

Figure 5-16. *Showing formatted confusion matrices for the salary prediction problem solved*

As you can see, the model does considerably well because Micro Accuracy is close to 1 and Log Loss is very small. The default generated code doesn't have the code to print the details of the confusion matrix. However, you can easily get to that by adding the following code:

```
var confusionMatrices = crossValResults.Select(r => r.Metrics.
ConfusionMatrix);
```

in
```
public static void PrintMulticlassClassificationFoldsAverageMetrics
(IEnumerable<TrainCatalogBase.CrossValidationResult<MulticlassClassificati
on
Metrics>> crossValResults)
```

All of these matrices are available as properties of `MulticlassClassification Metrics` as shown here:

```
⊞ Assembly Microsoft.ML.Data, Version=1.0.0.0, Culture=neutral, PublicKeyToken=cc7b13ffcd2ddd51

  using System.Collections.Generic;

⊟namespace Microsoft.ML.Data
  {
⊞     ... public sealed class MulticlassClassificationMetrics
      {
⊞         ... public double LogLoss { get; }
⊞         ... public double LogLossReduction { get; }
⊞         ... public double MacroAccuracy { get; }
⊞         ... public double MicroAccuracy { get; }
⊞         ... public double TopKAccuracy { get; }
⊞         ... public int TopKPredictionCount { get; }
⊞         ... public IReadOnlyList<double> PerClassLogLoss { get; }
⊞         ... public ConfusionMatrix ConfusionMatrix { get; }
      }
  }
```

ML.NET trainers for classification

ML.NET provides several trainers for binary and multiclass classifications.

Binary classifiers

Table 5-1 shows several binary classifiers available in ML.NET and where they are in the framework.

Table 5-1. *Binary classifiers and their location in the framework*

Classifier name	Encapsulated as
AvergePerceptron	BinaryClassification.Trainers.AveragedPerceptron
FieldAwareFactorization Machine	BinaryClassification.Trainers. FieldAwareFactorizationMachine
LbfgsLogisticRegression	BinaryClassification.Trainers. LbfgsLogisticRegression
LinearSvm	BinaryClassification.Trainers.LinearSvm
Prior	BinaryClassification.Trainers.Prior
SdcaLogisticRegression	BinaryClassification.Trainers. SdcaLogisticRegression
SdcaNonCalibrated	BinaryClassification.Trainers.SdcaNonCalibrated
SgdCalibrated	BinaryClassification.Trainers.SgdCalibrated
SgdNonCalibrated	BinaryClassification.Trainers.SgdNonCalibrated
FastTree	BinaryClassification.Trainers.FastTree
FastForest	BinaryClassification.Trainers.FastForest

Multiclass classifiers

Table 5-2 shows several multiclass classifiers available in ML.NET and where they are in the framework.

Table 5-2. *Multiclass classifiers and their location in the framework*

Classifier name	Encapsulated as
LbfgsMaximumEntropy	MulticlassClassification.Trainers. LbfgsMaximumEntropy
NaiveBayes	MulticlassClassification.Trainers.NaiveBayes
OneVersusAll	MulticlassClassification.Trainers.OneVersusAll
PairwiseCoupling	MulticlassClassification.Trainers. PairwiseCoupling
SdcaMaximumEntropy	MulticlassClassification.Trainers. SdcaMaximumEntropy
SdcaNonCalibrated	MulticlassClassification.Trainers. SdcaNonCalibrated

Setting up options for the classifier

If you take a close look at the generated code, you shall see that the arguments of OneVersusAll are set up like this. The code is pretty-printed here to make it more readable. The generated code is not pretty-printed.

```
var trainer = mlContext.MulticlassClassification.Trainers.OneVersusAll
(
mlContext.BinaryClassification.Trainers.FastTree
  (new FastTreeBinaryTrainer.Options()
        {
                NumberOfLeaves = 26,
                MinimumExampleCountPerLeaf = 1,
                NumberOfTrees = 20,
                LearningRate = 0.05887203f,
                Shrinkage = 3.070639f,
                LabelColumnName = "Salary",
            FeatureColumnName = "Features"
        }),

        labelColumnName: "Salary"
)
```

In this call, a multiclass classifier "OneVersusAll" is being configured. It takes two parameters. The first one is the binary classifier that it needs to use to differentiate one class from all other and the name of the label column (in this case "Salary").

The following screenshots show how you can explore different options that these classifiers can take (Figure 5-17).

```
var trainer = mlContext.MulticlassClassification.Trainers.SdcaMaximumEntropy()
```
▲ 1 of 2 ▼ (extension) Microsoft.ML.Trainers.SdcaMaximumEntropyMulticlassTrainer MulticlassClassificationCatalog.MulticlassClassificationTrainers.SdcaMaximumEntropy(**Microsoft.ML.Trainers.**SdcaMaximumEntropyMulticlassTrainer.Options **options**)
Create Microsoft.ML.Trainers.SdcaMaximumEntropyMulticlassTrainer with advanced options, which predicts a target using a maximum entropy classification model trained with a coordinate descent method.
options: *Trainer options.*

Figure 5-17. *Showing how to set up the trainer options for SdcaMaximumEntropy*

Normally, there are two overloads for most of the trainers. The first one takes an Option type which can store all the configuration values, and the other one generally allows to pass all the configurations as literals and numeric values.

Here are the two overloads of this particular trainer:

```
public static SdcaMaximumEntropyMulticlassTrainer SdcaMaximumEntropy(this
MulticlassClassificationCatalog.MulticlassClassificationTrainers catalog,
SdcaMaximumEntropyMulticlassTrainer.Options options);


public static SdcaMaximumEntropyMulticlassTrainer SdcaMaximumEntropy(this
MulticlassClassificationCatalog.MulticlassClassificationTrainers catalog,
string labelColumnName = "Label", string featureColumnName = "Features",
string exampleWeightColumnName = null, float? l2Regularization = null,
float? l1Regularization = null, int? maximumNumberOfIterations = null);
```

So we can configure such a trainer like this (this one uses the second overload)

```
var trainer = mlContext.MulticlassClassification.Trainers.
SdcaMaximumEntropy("Salary", "Features", null, 0.2334f, 0.454f, 100);
```

or like this

```
var trainer =
mlContext.MulticlassClassification.Trainers.SdcaMaximumEntropy
 (
new Microsoft.ML.Trainers.SdcaMaximumEntropyMulticlassTrainer.Options()
        {
```

```
        BiasLearningRate = 0.35f,
        ConvergenceCheckFrequency = null,
        ConvergenceTolerance = 0.23f,
        FeatureColumnName = "Features",
        ExampleWeightColumnName = string.Empty,
        L1Regularization = 0.12f,
        L2Regularization = 0.22f,
        LabelColumnName = "Salary"
    });
```

All other trainers can be configured this way.

Summary

In this chapter, you have learned how to use ML.NET model builder to your advantage to locate/discover the perfect or near perfect classifier for the dataset in question. You have also learned how to evaluate the model. The flow is always to prepare the data, feed that to model builder to locate/discover the best model, and then run several experiments on that model to make its performance better.

CHAPTER 6

Clustering

Birds of a feather flock together

Introduction

Sometimes, we prepare a long list of grocery items and go to the supermarket well prepared to buy what we want. However, sometimes the midday sugar trigger can send us to the supermarket like a dart for picking up a chocolate. People who prepare list of items generally spend way more time in the store than those who do not. From the perspective of the store buyers, spending more time in the store is a great thing. These buyers are what we can call "Organized Buyers". They know what they want and how much of it that they want. On the flip side, we have those buyers who just drop in the store for picking one or two items on a real physical/mental need trigger. These people are what we can call "Disorganized Buyers". Clustering is an unsupervised machine learning technique to automatically categorize datasets like these customers/buyers are for the store. In more general terms, clustering can be thought of as automatic grouping of things, behaviors, and so on. There is obviously a known right answer to the number of groups present in a dataset, but it is impossible to be known for each and every dataset in prior.

Objective of this chapter

In this chapter, you shall learn about a couple of algorithms to cluster/segregate a dataset into multiple clusters. The example we just discussed has two groups or clusters – one encompassing the organized buyers and other the disorganized ones. After reading this chapter, you shall have a thorough understanding of how popular clustering algorithms work and how to measure their performance using ML.NET.

© Sudipta Mukherjee 2021
S. Mukherjee, *ML.NET Revealed*, https://doi.org/10.1007/978-1-4842-6543-7_6

Intuition behind K-Means

One of the most popular algorithms used in clustering is *K-Means clustering* algorithm. It relies on the fact that anything can be represented as a vector in N-dimensional space. This may sound very complex at first, but it is not. It is just a special use of geometry. I shall explain it here.

In high school mathematics, you learned about coordinate geometry. Now we shall go through a geometrical interpretation of the clustering problem, and it's based on the theory you learned in coordinate geometry. Trust me!

As you can see in the following texts, each customer can be represented by a point in a two-dimensional plane where the X coordinate denotes the hours spent in the store and Y coordinate denotes the average number of items they purchased. So you can now see from the imaginary plot in Figure 6-1 that the dots at the lower corner denote buyers who spent less time in the store (low X value) and bought fewer items (low Y value). These are the people who the store wants to label as "Disorganized Buyers" – not on their face but in the store's database!

On the other extreme, we have people who spent more time in the store and bought more than average items for all other buyers. These are the people the store wants to label as "Organized Buyers".

The encompassing bubble around the points drawn as broken lines in circular shape denote the cluster/group. You can think of these bubbles as the border or outline of the cluster. If you like analogies, these are more like the boundary walls that protect the cluster inside. So any point that falls between this encompassing circle is thought to be a member of the cluster.

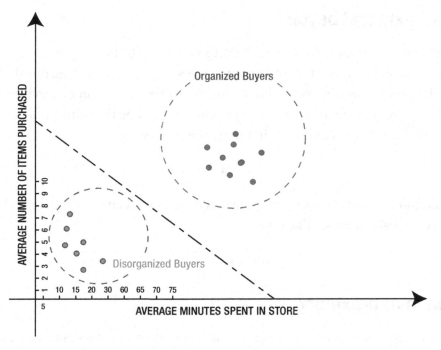

Figure 6-1. *Showing two clusters of customers in customer segmentation*

The main goal of a clustering algorithm is to move closely related data points (in the case of this example, the data of each customer) in the same cluster such that their distance (also known as *intra-cluster distance*) is as less as possible and the distance between clusters (also known as *inter-cluster distance*) is as big as possible. Later in the chapter, you shall see how to use several matrices to quantify the performance of a clustering algorithm.

A bit of mathematics

The act of segregating customers into several clusters is called *"Customer Segmentation"*, and it is a popular application of clustering algorithm. Moreover, it is easy to start with few dimensions of the data that makes sense already and gradually move to make the representation more detailed.

Each customer is represented in two dimensions as a point. Let us say we denote ith customer as Cust$i$.

Then, we can write

$$Cust_i = \left(x_i, y_i \right)$$

where x_i denotes the amount of time spent in the store and y_i denotes the number of items purchased.

The third axis and beyond

So far, we have represented data for each customer as a two-dimensional point where the two axes were the amount of time spent in the store and the average number of items purchased. However, as you can probably imagine, it could be extended to have more details. For example, the *number of visits per month* could be the third value if we want to represent the data for the customers in three dimensions.

$$Cust_i = (x_i, y_i, z_i)$$

Extrapolating on this, you can imagine that a customer can be represented by a m point in m-dimensional space like this:

$$Cust_i = (x_i, y_i, ...m_i)$$

The notion of proximity

Now that we have successfully represented each customer as a data point, we can find their distance between one another using Euclidean distance function that you learned in coordinate geometry. This distance will give us a sense of proximity between two data points (in this case two customers). This will be the clue needed to put two customers in close proximity in the same cluster or group. The following illustration (Figure 6-2) attempts to make a visualization.

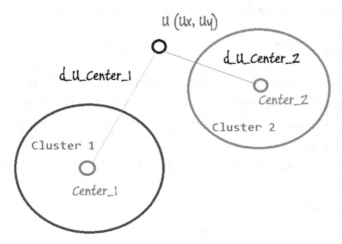

Figure 6-2. *Showing a couple of clusters and an unknown data point*

d_U_Center_1 is the Euclidean/Manhattan distance between the cluster centroid Center_1 (which is the cluster centroid for cluster 1) and the unknown data point.

d_U_Center_2 is the Euclidean/Manhattan distance between the cluster centroid Center_2 (which is the cluster centroid for cluster 2) and the unknown data point.

If d_U_Center_1 < d_U_Center_2, then the unknown data point should be attached to the first cluster as the centroid for that is nearer than the other centroid. Otherwise, it should be attached to the second cluster.

This process of cluster assignment is iterative, and it continues for the whole part of K-Means clustering where the initial guesses for the clusters shift and finally settle to become the final centroids of the clusters.

The distance metric used is generally the Euclidean distance or the Manhattan distance (a.k.a. city block distance). The following section provides a refresher for you for these distance matrices.

The Euclidean distance

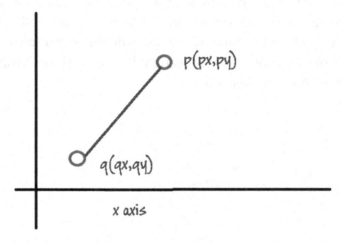

Figure 6-3. *Euclidean distance*

If there are two points denoted by $p(x, y)$ and $q(x, y)$, Euclidean distance between these two points is represented by the following formula:

$$D_{p,q} = \sqrt{\left(q_x - p_x\right)^2 + \left(q_y - p_y\right)^2}$$

Here, p_x denotes the value of x coordinate for the point p and so on.

Euclidean distance in more dimension

As you can see, extrapolating on the previous equation, we can get the general equation for calculating Euclidean distance between two points in m dimension as follows:

$$D_{p,q} = \sqrt{\left(q_x - p_x\right)^2 + \left(q_y - p_y\right)^2 + \ldots + \left(q_m - p_m\right)^2}$$

Sometimes it could be needed to use other distances metrics like Manhattan or city block distance because calculating Euclidean distance can be computationally expensive. Here is the equation for Manhattan or city block distance:

$$d_{cityblock}\left(\mathbf{p} \cdot \mathbf{q}\right) = \sum_{i=1}^{n} |p_i - q_i|$$

Centroid, the center of the cluster

Centroid in mathematics and physics denotes the point which is the mean of all the points on a given shape of any contour. Figure 6-4 shows the calculation formula for a centroid of a concave polygonal shape. The shape is deliberately drawn like this because in the real life example data points can be scattered like this. The centroid is the mean of all the coordinates as shown in Figure 6-4.

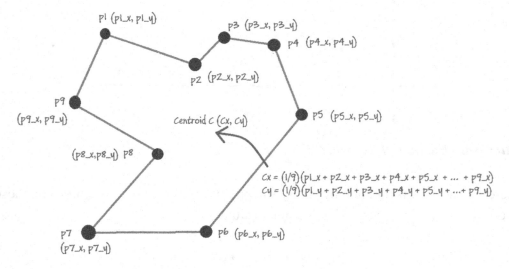

Figure 6-4. *Explaining centroid*

So in m-dimensional place, the centroid can be represented by the following formulae, where C_x denotes the X coordinate value and so on and C_m denotes the value of the centroid at *m*th coordinate:

$$C_x = \frac{1}{N} \cdot \sum_{i=0}^{N} x_i$$

$$C_y = \frac{1}{N} \cdot \sum_{i=0}^{N} y_i$$

$$C_m = \frac{1}{N} \cdot \sum_{i=0}^{N} m_i$$

So you can see that centroid coordinates are nothing but the mean or average of the projected coordinates of all the points on the shape at a given axis. The following C# function finds the centroid:

```
List<double> CentroidLocations(List<List<double>> points)
=> Enumerable.Range(0,points[0].Count)
  .Select(z =>  points.Select(p =>p[z]).ToList())
  .Select(z => z.Average())
  .ToList();
```

The following function calculates Euclidean distance between any two points represented in N dimension:

```
double EuclideanDistance(List<double> p, List<double> q)
  => Math.Sqrt(p.Zip(q,(px,qx) =>
              Math.Pow(px - qx,2)).Sum());
```

Here is a client code that uses these two functions to make some of the points, made thus far, more clear for you:

```
void Main()
{
            var p = new List<double>(){1,2,3};
            var q = new List<double>() { 1, 3, 4 };
            var r = new List<double>() { 1, 2, 31 };
```

```
        var s = new List<double>() { 1, 3, 41 };
        var t = new List<double>() { 11, 2, 3 };
        var u = new List<double>() { 1, 31, 4 };
        var centroid = CentroidLocations(new List<List<double>>()
        {
                p, q, r, s, t
        });
        var distance_centroid_p = EuclideanDistance(centroid,p);
        var distance_centroid_q = EuclideanDistance(centroid,q);
        Console.WriteLine($"distance p and centroid =
                {distance_centroid_p}");
        Console.WriteLine($"distance q and centroid =
                {distance_centroid_q}");
}
```

Although you shall use ML.NET to do clustering, it is always good to know the details of the internals.

Keep moving the centroid until it's not moving much

The reason it is called K-Means because it takes an initial guess of K for the number of clusters. These settled centroids are then declared as the cluster centroids and count of such centroids becomes the count of the clusters available in the dataset.

The algorithm has these major steps:

1. Initialization (guessing the initial centroid locations).

2. Update centroid at each step (shifting the centroids as calculated per iteration).

3. Verify if there is need to continue; if not, stop and report the clusters.

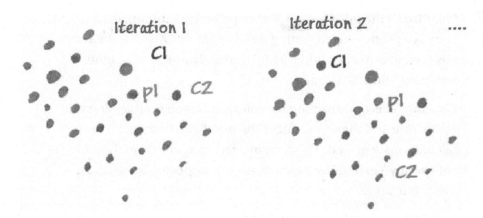

Figure 6-5. *Showing shifting centroids in an imaginary dataset*

Figure 6-5 attempts to show how the centroids move in a step in the iterative process of K-Means clustering. The image shows only a couple of imaginary iterations. In a real case obviously, there will be more iterations. In ML.NET K-Means option, you can set the number of maximum iterations by setting

`KMeansTrainer.Options.MaximumNumberOfIterations`

Note that point P1 initially belonged to the first cluster (with centroid located at C1), but later due to the shift of the cluster coordinates belonged to the second cluster (for which centroid is located at C2).

Also K-Means needs the number of clusters as a parameter, and unsurprisingly you can find that too in the K-Means option provided by ML.NET at `KMeansTrainer. Options.NumberOfClusters`.

Initialization

There are several kinds of initialization possible in K-Means clustering. You can either choose any random points as the initial clusters or use either of a couple of optimizations available to make a smarter guess of the initial clusters.

The three variations of initialization provided by ML.NET are

1. **PlusPlus** (this is an implementation of K-Means++ algorithm)

2. **Yinyang** (this is an implementation of the Yinyang optimization)

3. **Random** (random points are picked as centroids)

PlusPlus: If the initial centroids are picked at random, then there is a probability of getting poorly defined clusters which are way too close to each other. This is a disadvantage of randomly assigning initial centroids.

Yinyang: This optimization technique relies on double filters except one and thereby reduces the number of distance calculations required. This scheme gets its name from the Chinese philosophy Yin and Yang, which are two opposing forces that create harmony.

Random: This is the naïve option to choose centroids at random.

All these initialization strategies are available as option of K-Means trainer as KMeansTrainer.InitializationAlgorithm.

Table 6-1. *Showing options to initialize K-Means via ML.NET options*

Algorithm	Encapsulated as
K-Means++	**KMeansTrainer.InitializationAlgorithm.KMeansPlusPlus**
Yinyang	**KMeansTrainer.InitializationAlgorithm.KMeansYinyang**
Random	**KMeansTrainer.InitializationAlgorithm.Random**

Update of centroids

At each step of the iterative process, the centroids associated with the points get changed. Gradually toward the end, when the algorithm converges, the centroids become more still; in other words, their coordinates do not shift much anymore.

Clustering Iris flowers using ML.NET

This is the easy part. The reason is as follows:

Step 1: Create a new .NET core console application (Figure 6-6).

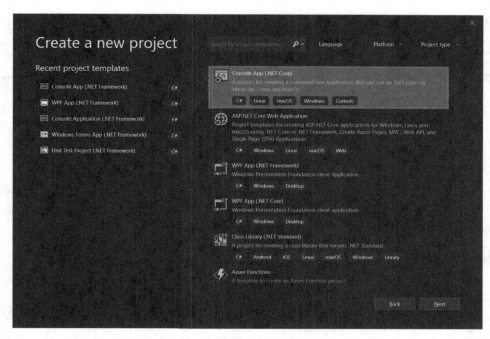

Figure 6-6. *Showing the console application*

Step 2: Configure the project.

Figure 6-7. *Provide a name*

Step 3: Get the NuGet Package.

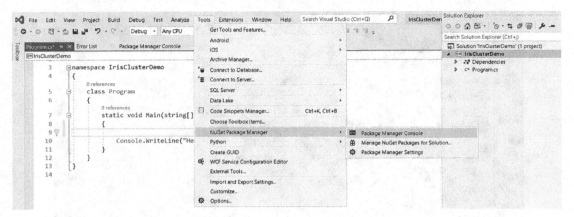

Step 4: Get the package.

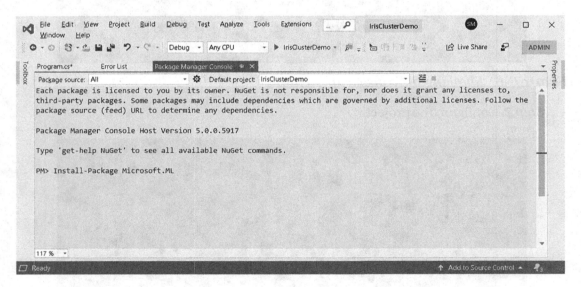

Step 5: After the NuGet is installed.

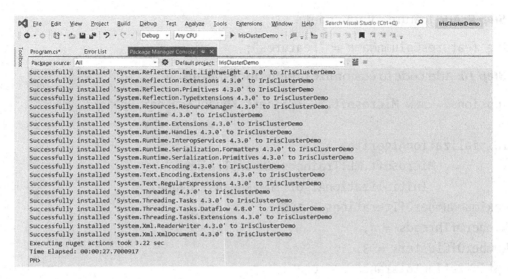

Figure 6-8. *Showing the successful installation of ML.NET NuGet Package*

Step 6: Once the package is installed, the IntelliSense will figure out the paths.

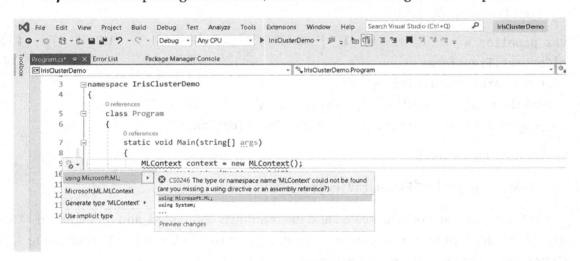

Step 7: Add the path to the Iris flower dataset file.

```
string _dataPath = @"C:\Users\Sudipta\Downloads\iris.data";
```

Step 8: Add code to read the Iris dataset in a IDataView instance.

```
IDataView dataView = mlContext.Data.LoadFromTextFile<IrisData>
(_dataPath, hasHeader: true, separatorChar: ',');
```

Step 9: Add feature column name.

```
string featuresColumnName = "Features";
```

Step 10: Add code to customize options for K-Means.

```
var options = new Microsoft.ML.Trainers.KMeansTrainer.Options
{
    InitializationAlgorithm =
              Microsoft.ML.Trainers.KMeansTrainer.
                InitializationAlgorithm.KMeansYinyang,
    MaximumNumberOfIterations = 100,
    NumberOfThreads = 4,
    NumberOfClusters = 3,
    OptimizationTolerance = 0.002F,
    FeatureColumnName = featuresColumnName
};
```

Step 11: Create the pipeline.

```
var pipeline =
mlContext.Transforms
.Concatenate(featuresColumnName,
 "SepalLength", "SepalWidth", "PetalLength", "PetalWidth")
.Append(mlContext.Clustering.Trainers.KMeans(options));
```

Step 12: Fit the model.

```
var model = pipeline.Fit(dataView);
```

At this point, the model is created and the training is completed, and now you can use this model to predict what cluster a new data point should belong to. The following steps show how to predict that for a given data point:

Step 13: Create an example instance.

```
IrisData Setosa = new IrisData
          {
                SepalLength = 5.1f,
                SepalWidth = 3.5f,
```

```
            PetalLength = 1.4f,
            PetalWidth = 0.2f
        };
```

Step 14: Create the prediction engine from the model and perform the prediction.

```
var predictor = mlContext.Model
.CreatePredictionEngine<IrisData, ClusterPrediction>(model);

var prediction = predictor.Predict(Setosa);

Console.WriteLine($"Cluster:{prediction.PredictedClusterId}");
Console.WriteLine($"Distances: {string.Join(" ",
prediction.Distances)}");
```

This prints the following output:

Cluster: 2
Distances: 16.87281 0.03447342 0.630455

The distances denote the distances of this given data point from the calculated centroids of the clusters. As the distance from the second cluster is minimal, thus the algorithm determines that the given data point should belong to this cluster.

Getting centroid locations

To get the locations of the centroid, add the following code:

```
VBuffer<float>[] centroids = default;

var modelParams = model.LastTransformer.Model;
modelParams.GetClusterCentroids(ref centroids, out int k);

//Printing coordinates of the centroids
for (int i = 0; i < centroids.Length; i++)
{
  Console.WriteLine($"Centroid #{i + 1} is located at " +
                $@"({centroids[i].GetValues()
                          .ToArray()
```

```
                    .Select(t => t.ToString("F2"))
            .Aggregate((f, s) => f + "," + s)})");

}
```

For this Iris flowers, we get the following result:

Cluster: 1
Distances: 0.02159119 11.69127 25.59897
Centroid #1 is located at (5.01,3.42,1.46,0.24)
Centroid #2 is located at (5.90,2.75,4.39,1.43)
Centroid #3 is located at (6.85,3.07,5.74,2.07)

At a second run, you can get the following result (this may vary at your end):

Cluster: 2
Distances: 11.64523 0.02159119 25.31428
Centroid #1 is located at (5.88,2.74,4.39,1.43)
Centroid #2 is located at (5.01,3.42,1.46,0.24)
Centroid #3 is located at (6.85,3.08,5.72,2.05)

That is because the MLContext is created without a seed, and therefore the results obtained are nondeterministic. If you want to make sure that you get the same result from the machine learning pipelines in ML.NET, then create MLContext with a seed like this (as shown in Figure 6-9).

```
MLContext mlContext = new MLContext(seed : 1);
```

Figure 6-9. *Showing declaration of MLContext with a seed to ensure deterministic behavior*

After the initialization is set to seed : 1, then a couple of runs produce the following output, and as you can run the script as many times as you would like and every time, it will be exactly the same.

Cluster: 1
Distances: 0.02159119 25.59896 11.69127
Centroid #1 is located at (5.01,3.42,1.46,0.24)
Centroid #2 is located at (6.85,3.07,5.74,2.07)
Centroid #3 is located at (5.90,2.75,4.39,1.43)

Validating the model with ground truths

In the current setting of the experiment, we have an undue advantage, which for the most part won't be present in real-life clustering experiments. We have the original labels or the cluster information each flower should belong to; and we can use this information to validate how the clustering model worked. The more number of flowers belonged to the right cluster, the better.

In the dataset, the tags "Iris-Setosa", "Iris-Versicolor", and "Iris-Virginica" occur in succession. And if you use deterministic model, then the clusters would be placed accordingly. In other words, the first cluster will denote cluster of "Iris-Setosa" flowers and so on.

Add the following code to check ground truth for the model:

```
//Ground truth verification
string[] labels = new string[]
{ "Iris-setosa","Iris-versicolor", "Iris-virginica" };

var sepalLengths = dataView.GetColumn<float>("SepalLength");
var petalLengths = dataView.GetColumn<float>("PetalLength");
var sepalWidths  = dataView.GetColumn<float>("SepalWidth");
var petalWidths  = dataView.GetColumn<float>("PetalWidth");
Func<string, int> toIndex = p => Array.IndexOf(labels, p) + 1;

var groundTruths = File.ReadAllLines(@"iris.data")
                    .Skip(1)//Skip header
                    .Select(t => toIndex( t.Split(',')[4]));

int count = 0;
for (int index = 0; index < sepalLengths.Count(); index++)
{
      IrisData temp = new IrisData
      {
                  SepalLength = sepalLengths.ElementAt(index),
                  SepalWidth = sepalWidths.ElementAt(index),
                  PetalLength = petalLengths.ElementAt(index),
                  PetalWidth = petalWidths.ElementAt(index)
      };
```

```
var predicted = predictor.Predict(temp);
//Ground truth check
if (predicted.PredictedClusterId !=
        groundTruths.ElementAt(index))
                count++;
}

var totalRows = sepalLengths.Count();
double correctlyClustered =
  Math.Round( 100*(double)(totalRows - count) / totalRows,2);

Console.WriteLine($"{correctlyClustered}% belong to the right cluster");
```

When run with KMeansPlusPlus strategy, Iris dataset proved to be very nicely clustered. Using KMeansPlusPlus, 89.33% records were correctly clustered. For random initialization, only 2% were rightly clustered. So you can see the initialization has a drastic effect on the performance of the clustering algorithm.

Evaluating the model in the wild

In the absence of ground truth, the model is generally evaluated on two factors:

1. How densely the elements/data points in the cluster are packed.

2. How far different clusters are from one another?

ML.NET provides these three metrics in the ClusteringMetrics class.

Average distance (AD)

Average score. For the K-Means algorithm, the "score" is the distance from the centroid to the example. The average score is, therefore, a measure of proximity of the examples to cluster centroids. In other words, it is a measure of "cluster tightness". Note, however, that this metric will only decrease if the number of clusters is increased, and in the extreme case (where each distinct example is its own cluster), it will be equal to zero. The lower this distance is, the better – depicting closely knit clusters.

Davies-Bouldin index (DBI)

Davies-Bouldin index is a measure of how much scatter is in the cluster and the cluster separation. The higher the number, the better – representing the cluster centroids are really scattered far from each other.

Normalized mutual information (NMI)

Normalized mutual information is a measure of the mutual dependence of the variables. This metric is only calculated if the Label column is provided.

For the current example strategies, these are calculated along with time to complete the clustering process. These details are captured in Table 6-2.

To get these data about performance, use the following code:

```
ClusteringMetrics metrics =  mlContext.Clustering.Evaluate(model.
Transform(dataView)
,"PredictedLabel", "Score", "Features");

Console.WriteLine(metrics.AverageDistance);
Console.WriteLine(metrics.DaviesBouldinIndex);
Console.WriteLine(metrics.NormalizedMutualInformation);
```

Table 6-2. *Results*

Strategy	Metrics	Time taken (100 iterations)
Random	0.968531201680501(AD) 0.952212697058903 (DBI) 1 (NMI)	525 ms
Yinyang	0.526269976298014 0.662323100264084 1	522 ms
PlusPlus	0.526271146138509 0.662323100264084 1	471 ms

As you can see from the table, **KMeansPlusPlus** is the fastest and also probably the best choice you can make because it performed the best in ground truth validation also.

Summary

You have learned how K-Means works and how to measure its performance. Sometimes when the ground truth labels are known, it may seem that clustering is basically classification. But clustering is generally done with datasets for which ground truth labels are not known. And it may make sense to realize that there is not an exact right answer to a clustering problem, but a close-enough approximation to validate any hypothesis is sufficient as an outcome of a clustering experiment. For example, the store might guess that there are three types of buyers and a clustering experiment is required to validate this hypothesis and then the store can find those customers and give away customized offers which will lure them more than a blanket one. However, there are other kinds of clustering algorithm like density-based clustering algorithm DBSCAN, which may outperform K-Means because K-Means for the most part starts with an unassuming set of clusters (even with KmeansPlusPlus). The disadvantage of K-Means is that you have to supply the value of K. But domain knowledge or hypothesis can help supply an initial guess.

CHAPTER 7

Sentiment Analysis

Are you happy or not, that's the question!

Introduction

Sentiment about a product or a service offered by a company is all too valuable in this era than ever before. Knowing whether their established customer base and potential customers are showing a positive or negative sentiment (as shown in Figure 7-1) toward their product or service can be game-changing for companies. However, extracting the true sentiment from a phrase written in English is a challenge, let alone in all the languages. That's because human languages could be ambiguous, and we can be sarcastic at time and understanding sarcasm is a huge challenge for computers. Although we are getting better with each passing year, but it is still a long way to go.

Figure 7-1. *Interpreting sentiment*

© Sudipta Mukherjee 2021
S. Mukherjee, *ML.NET Revealed*, https://doi.org/10.1007/978-1-4842-6543-7_7

In this chapter, you shall see how ML.NET can help you to do sentiment analysis from textual sources of data. So ideally after finishing this chapter, you should be able to perform sentiment analysis tasks (bipolar, either positive or negative) on your data in your domain.

Basic ideas

There are two basic approaches to solving sentiment analysis tasks. The following section walks you through these two and shows the pros and cons of both approaches.

Consider these two statements:

"The plot of the movie was truly *unpredictable*".

"The steering wheel of this new car is rather *unpredictable*".

A movie plot being *unpredictable* makes it *desirable*. A car with *unpredictable* steering wheel makes it *dangerous* at best. So the first statement shows a positive sentiment about a new play, while the second one displays a really stark negative sentiment about the new car.

First idea

Sentiment analysis works with a simple algorithm. The main idea is simple. Every word either expresses a positive or a negative sentiment for each domain. For example, the word "unpredictable" may be rated to have a higher positive sentiment while being used in movie reviews. The same word has to have a high negative sentiment score when used in the context for car reviews.

The idea to calculate overall sentiment score for a given phrase/sentence is simple. All you have to do is to keep adding the positive and negative scores of each of the tokens/words from the phrase/sentence.

If the overall positive score is more than the overall negative score, then we claim that the phrase/sentence in question might be expressing a positive attitude. On the contrary, if the overall negative score beats out overall positive score, then we can conclude that the sentence/phrase probably expresses a negative sentiment.

The words are called "*Lexicons*" in this context.

For each lexicon/word, there is a positive score and a negative score representing its expressed positive and negative sentiment. These numbers are available in a few specialized dictionary-like structures. One such structure is SentiWordNet.

You can download it from

SentiWordNet - Cnr

sentiwordnet.isti.cnr.it/ ▼

Human Language Technology Group. This is the homepage of the Human Language Technology group of NeMIS-ISTI-CNR. **SentiWordNet**.

One entry in this dictionary looks like this:

```
"able#" "0.125" 0
```

where "able" is the word or lexicon, and it expresses a positive sentiment score of 0.125 and it is generally not used in negative sentiment; thus, the score for negative sentiment score is zero.

The following code helps to read and create an in-memory representation of the SentiWordNet dictionary:

```
void Main()
{
        var sentiWordList = System.IO.File.ReadAllLines
        (@"SentiWordNet_3.0.0.txt")
.Where(line => !line.StartsWith("#"))
.Select(line => line.Split('\t'))
.Where(tokens => tokens.Length >= 5)
.Select(lineTokens => new
                {
                    POS = lineTokens[0],
                    ID = lineTokens[1],
                    PositiveScore = lineTokens[2].Trim(),
                    NegativeScore = lineTokens[3].Trim(),
                    Words = lineTokens[4]
                })
.Select(item => new string[]
        {
                item.Words.Substring(0, item.Words.LastIndexOf('#')
                + 1),
                 item.PositiveScore,
                item.NegativeScore
        });
```

115

```
        foreach (var element in sentiWordList.Take(5))
        {
          //The following line should be in a single line
              Console.WriteLine($@"{element.Lexicon}
          {element.PositiveScoe} {element.NegativeScore}");
        }
```

This produces the following output:

```
able#  0.125 0
unable#  0 0.75
dorsal#2 abaxial#  0 0
ventral#2 adaxial#  0 0
acroscopic#  0 0
```

POS stands for Parts of Speech.

The following function gets the polarity (positive and negative sentiment expressed by a given word) from the sentiment dictionary:

```
private Tuple<float, float> GetPolarity(IEnumerable<string[]>
sentiWordNetList, string word)
{
        var matchedItem = sentiWordNetList
         .FirstOrDefault(item => item.ElementAt(0).Contains(word));
        if (matchedItem != null)
        {
                return new Tuple<float,
                float>(Convert.ToSingle(matchedItem[1]),//positive
                 Convert.ToSingle(matchedItem[2]));//negative
        }
        else
                return new Tuple<float, float>(OF, OF);
}
```

This produces the following when called for the word "good":

A tuple with 0.625 and 0

Extrapolating on this method, the following method calculates the polarity (either positive or negative) of a complete sentence:

```
private int GetPolarityScore(string sentence, IEnumerable<string[]>
sentiWordNetList)
{
        var words = sentence.Split (' ');

var polarities = words.Select( word => GetPolarity (sentiWordNetList,
word));

var totalPositivity = polarities.Sum(p => p.Item1);

var totalNegativity = polarities.Sum(p => p.Item2);

Console.WriteLine($"Positive polarity of this sentence is
{totalPositivity}");
Console.WriteLine($"Negative polarity of this sentence is
{totalNegativity}");

if (totalPositivity > totalNegativity) return 1;

else if( totalNegativity == totalPositivity) return 0;
else return -1;

}
```

When called with the following arguments,

**GetPolarityScore("I love this awesome product I thought the camera will be
great much better though", sentiWordList)**

it returns 1 and prints the following lines about polarity:

Positive polarity of this sentence is 3.5 Negative polarity of this sentence is 1.5

However, there is a caveat with this approach. It might not be easily evident, but if you think about it, it becomes really easy to spot. That's about handling negations in input phrases.

Handling negations

Sometimes, we use negative words to describe positive mood and vice versa, like the two example sentences in Figure 7-2.

"The angle of the camera was not good"

This one echoes a *negative* sentiment

"The angle of the camera was not bad"

This one echoes an *Okish* (almost positive) sentiment

Figure 7-2. *Words explain moods*

The idea to get past these is to create two set of combos by extracting positive and negative words from SentiWordNet. A positive word is one for which the positive sentiment score is more than the negative sentiment score. A negative word is a word for which the negative sentiment score is more than the positive sentiment score.

Then, create two pairs of combos. One with positive words and negations and another with negations and negative words. The first pair of such combos will help extract cases that are truly bad (like "not good"), and the second pairs of such combos will help extract cases that are not as bad (like "not bad").

And here is the list of negations you can use:

- No
- Not
- Never
- Seldom
- Neither
- Nor

Generalization of sentiment analysis…

Another idea is to view sentiment analysis as a classic binary classification problem, where "Positive" and "Negative" are two classes, as seen in Figure 7-3.

```
AFRAID

AMUSED

ANGRY

HAPPY

ANNOYED

INSPIRED

SAD
```

Figure 7-3. *Positive and negative classes*

However, describing the sentiment analysis problem as a classical classification problem has its advantage. When expressed this way, the problem can be extended to extract sentiments that are beyond bipolar sentiments (positive/negative). Analyzing sentiment (not just polarity) is identifying the true sentiment like "Happiness", "Ecstasy", "Sadness", "Grumpiness", "Arrogance", "Anger", and "Indifference", to name a few.

When the sentiment analysis is capable of extracting emotions from data sources (textual, visual, videos), then it is called "*Emotion Detection*" or "*Emotion Analysis*".

Expressing sentiment analysis problem as a classic classification problem makes it easy to present itself as a supervised learning problem of classification of emotions (not binary but more). Imagine if you have labeled data from several people where their feeling/emotion/sentiment is tagged based on few input feature, it will be easy to feed this data to a supervised learning algorithm to get the predicted label of a newly arrived dataset entry.

Step 1: Create a console project in Visual Studio 2019. The community edition is free and can be downloaded from here `https://visualstudio.microsoft.com/vs/community/`.

Step 2: Right-click to add Machine Learning to this project.

Step 3: Select "Text classification" for sentiment analysis task as seen in Figure 7-4.

119

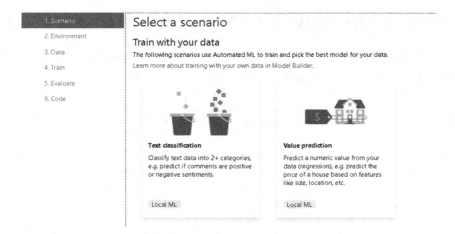

Figure 7-4. *Select a scenario*

This is because sentiment analysis is basically a text classification.

Step 4: Select the data file to start the training as seen in Figure 7-5.

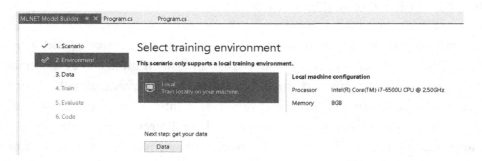

Figure 7-5. *Select the training environment*

At the time of this writing, this is supported only on local.

Data to be used in this application is from Sentiment140.

`http://help.sentiment140.com/for-students`

You can get the data from Google Drive by following the links on this URL.

Here are a few points about the data (taken from the preceding website).

The data is a CSV with emoticons removed. The data file format has six fields:

> **0**: The polarity of the tweet (0 = negative, 2 = neutral, 4 = positive).

> **1**: The ID of the tweet (2087).

> **2**: The date of the tweet (Sat May 16 23:58:44 UTC 2009).

3: The query (lyx). If there is no query, then this value is NO_QUERY.

4: The user that tweeted (robotickilldozr).

5: The text of the tweet (Lyx is cool).

Once the file is located, then the details are listed as follows.

In this information, the first column "**col0**" denotes the label of the sentiment analysis task. Remember that 0 indicates negative and 4 indicates positive. The last column "col5" denotes the text of the tweet. When the header row is skipped, then the program produces these autogenerated column names as seen in Figure 7-6.

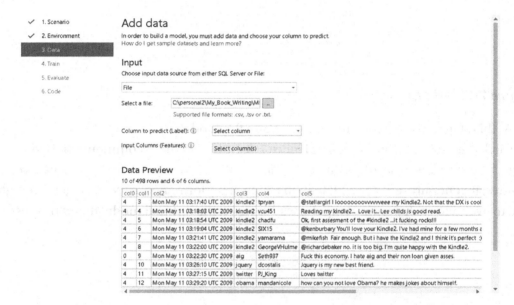

Figure 7-6. *Autogenerated column names*

Note   You may notice that we are not including the rest of the columns. Keep in mind that it is the task of a data scientist to choose which data input to use in the model.

Select the col0 as the data label column and the col5 as the input column as shown in Figure 7-7.

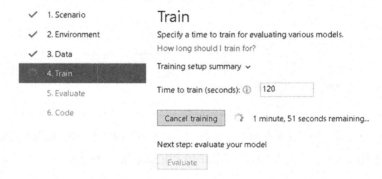

Figure 7-7. *Add the data*

At this stage, we are ready to train to obtain the model.

Click the Train button as seen in Figure 7-8. Although the recommended time range for training is in seconds range, my observation is that if you let it run a bit over 2 minutes, it generally reaches a plateau in terms of performance, because it gets enough time to evaluate each model along the way to reach the optimal model that it proposes.

Figure 7-8. *The training for the sentiment analysis model in progress*

Note   Clicking the link "How long should I train for?" will take you to the doc explaining the time needed depending on the size of the dataset.

At the end of the training, the wizard shows the training for the sentiment analysis model finalizing as seen in Figure 7-9.

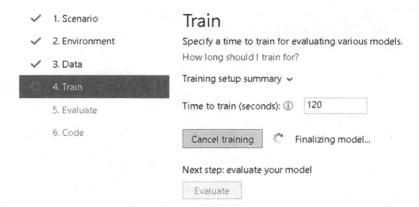

Figure 7-9. *The training for the sentiment analysis model in finalizing mode*

Step 5: The result of running the training for exactly 2 minutes produces this, as seen in Figure 7-10.

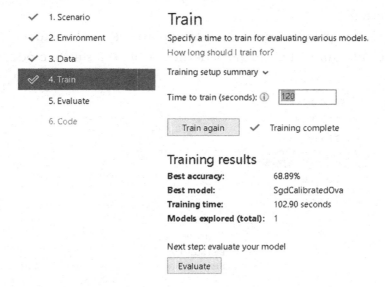

Figure 7-10. *Training complete*

And there you have it.

For comparison purposes, Figure 7-11 shows the result of running the training for 10 seconds.

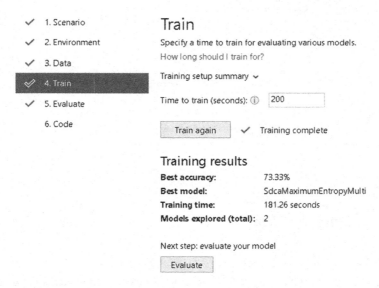

Figure 7-11. *Training results after 10 seconds*

As you can see, there is an obvious improvement in performance for training longer periods. But be warned that it will be hitting a plateau unless some other modifications are made to the data.

For the final attempt for training on this example dataset, I chose to run it for 200 seconds, as seen in Figure 7-12, and the performance of the resultant model is even better.

Train

Specify a time to train for evaluating various models.

How long should I train for?

Training setup summary ⌄

Time to train (seconds): ⓘ 200

Train again   ✓ Training complete

Training results

Best accuracy:	73.33%
Best model:	SdcaMaximumEntropyMulti
Training time:	181.26 seconds
Models explored (total):	2

Next step: evaluate your model

Evaluate

Figure 7-12. *Training results after 200 seconds*

The Model Builder wizard allows us to try the trained model via a generated UI (that is generated from the data on-the-fly). The following screenshot, Figure 7-13, shows this UI.

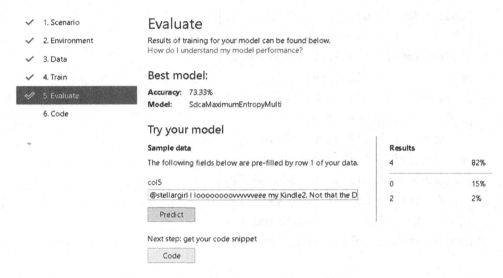

Figure 7-13. *Showing the result of the trained model*

You can add this generated model as the starting point to enhance the model. To add the generated code for the trained model, hit the "Add Projects" button as seen in Figure 7-14.

Figure 7-14. *Add Projects button*

Once the generated projects are added, it will have the following types:

```
public class ModelInput
{
        [ColumnName("col0"), LoadColumn(0)]
        public string Col0 { get; set; }
        [ColumnName("col1"), LoadColumn(1)]
        public float Col1 { get; set; }
        [ColumnName("col2"), LoadColumn(2)]
        public string Col2 { get; set; }
        [ColumnName("col3"), LoadColumn(3)]
        public string Col3 { get; set; }
        [ColumnName("col4"), LoadColumn(4)]
        public string Col4 { get; set; }
        [ColumnName("col5"), LoadColumn(5)]
        public string Col5 { get; set; }
}

public class ModelOutput
{
    // ColumnName attribute is used to change the column name from
    // its default value, which is the name of the field.
    [ColumnName("PredictedLabel")]
    public String Prediction { get; set; }
    public float[] Score { get; set; }
}
```

Then, this model can obviously be saved, loaded, and consumed to predict the sentiment of the new-coming entry like this:

```
var sampleData = new ModelInput()
{
  Col5 = @"@stellargirl I looooooooovvvvvveee my Kindle2.
      Not that the DX is cool, but the 2 is fantastic in its own right.",
 };

// Make a single prediction on the sample data and print results
var predictionResult = ConsumeModel.Predict(sampleData);
```

As you can probably guess, this particular text in question reflects a positive emotion. Therefore, the value will be 4.

Summary

The infrastructure provided by ML.NET allows you to do sentiment analysis as a special case of text analysis, but as you saw in the chapter, it is not that trivial and handling negations was just one of the caveats. As of this writing, ML.NET is continually evolving, and I hope to see a more in-depth categorization of feelings and emotions that extend beyond positive and negative for sentiment analysis. Although, you can use a deep learning model specially trained to do that and that is a workaround for now.

CHAPTER 8

Product Recommendation

You might be interested in this movie

Introduction

It's highly likely for individuals to look at similar product while shopping before purchasing what captures their imagination. Product recommendation is highly useful because it boosts sales. A few examples of recommendations are as follows:

© Sudipta Mukherjee 2021
S. Mukherjee, *ML.NET Revealed*, https://doi.org/10.1007/978-1-4842-6543-7_8

- Netflix recommends movies you might like based on what you watched and rated thus far.

- Spotify recommends music/songs based on users' preferences.

- Amazon recommends products that you can possibly be interested.

- Visual Studio IntelliCode offers code completion candidates based on previous examples written by other developers across different projects.

- Microsoft PowerPoint offers several design ideas based on the content. This design decision is the output of a recommender system that learns what appeals to users over time based on historical input and output.

These are just some examples of where a recommender system hide itself behind carefully crafted UIs, which sometimes is indistinguishable from magic. Recommender Systems is a very useful, and fortunately not so difficult to understand, application of supervised machine learning.

This chapter will introduce you to some key ideas of implementing recommender system and in particular will walk you through an example of product recommendation using a very popular algorithm called "matrix factorization". Along the way, you shall discover how ML.NET hides the complexities of such a system from the application developer.

Jargons of the trade...

Recommender systems recommend (quite obviously) something to people to *consider* (potential dates), *ponder* (potential job offers), *buy* (for purchasing stocks), *listen* (to songs the system thinks they would love), *watch* (movies the system believes they would resonate with), *accept a suggestion* (to change a phrase to what the recommender system thinks would sound more professional), and so on.

In all these contexts, a system is offering some help to people. The audience of recommender systems is referred to as Users in technical literature. And whatever the recommender system offers/recommends is called as items.

Users

Users are denoted by the word U, and for denoting m users, the subscript notation is used like this $U_i.... . U_m$.

Items

Items are denoted by the word I, and for denoting n items, the subscript notation is used like this $I_i.... . I_m$.

Ratings

Users rate items based on their experience, and then these ratings become available for collaborative filtering technique, predictions that are made based on existing ratings from others who have ratings in common with the active user to recommend an item based on previous preferences. Ratings given by users are generally denoted by $R_i ... R_n$.

So you can imagine that a previously rated item gallery would look like this:

Users	Items	Ratings	... Other Data
U1	I_1	3.5	
U2	I_2	?	
U2	I_1	5	

The challenge of the product recommendation algorithm is to fill the missing blocks like in this case the rating for second item is missing from the second user.

Type of recommender systems...

There are two major ways recommender systems are designed. The first set is called CBF (content-based filtering), and the second family of algorithms is called CF (collaborative filtering).

CBF relies on the fact that people generally like similar products, and this type of technique helps locate similar items or users who share the same preferences and then these information can be used to find the missing ratings of items.

However, in this chapter, we shall discuss about a popular algorithm called *matrix factorization* for performing collaborative filtering, a means of making automated predictions.

Normally in a recommender system, the number of users is way more (generally in the range of millions) and the number of items is less (generally in the range of thousands), but the total number of ratings is very sparse because not all users have rated all items.

Matrix factorization

Factorization means breaking a big number into two or more smaller numbers or expressions. Matrix factorization is a process to break a big matrix into two smaller matrices. In our case, we will represent the big matrix as a product of two smaller matrices. The first smaller matrix is the matrix between customers and their preferences for movies, and the second smaller matrix is the movies and their features (how comic the movie is, how much action is there in the movie, etc.). Elements in the bigger matrix are the dot product of a row and column of these two smaller matrices. The following table depicts it well.

Persons	Movie1	Movie2	Movie3	Movie4	Movie5
Dana	3	1	1	3	1
Ana	1	2	4	2	3
Sam	3	?	4	3	1
Hans	4	3	?	4	4

The size of this matrix is $m \times n$, where m is the number of users and n is the number of movies.

This is what the whole table will look if we store all the data in a single matrix. But storing the data this way won't be efficient because the matrix will have very large dimension (use a lot of machine memory), and at the same time, many elements in that matrix will be zero, because users didn't eat in every single movie. When a matrix has many zero elements, it is called a sparse matrix.

The two question marks on the table denote two ratings that we need to predict. What do you think Sam will rate "Movie2"? Looking at the table, it seems like Sam's preferences really are a close match to that of Dana's. So probably he will also hate the movie. So we can we can mark it as 1, Sam would not like the movie. Similarly, Han's expected Movie3 rating will be close to 4 because Han's preference is almost the same as Sam's. This approach of adjusting the predicted rating based on preferences of similar neighbors is called "*Collaborative Filtering*". It can be assumed as a process where all the similar neighbors (who have similar preferences) collaborate to filter out weeds and help produce the best possible recommendations for the new user.

Each element in this matrix is calculated from the dot product between the row and column of those two matrices. The first matrix is the preferences of the users. Users have preferences of several movie genres. Some users like comedy movies, some like action movies, few like thrillers, and if you are anything like me, you also like documentaries and mystery thrillers. So the preferences of users can be encoded by this binary representation as shown in the following matrix:

Persons	**Comedy**	**Action**	**Documentary**	**Thrillers**	**Biopic**
Dana	*Yes*	*No*	*No*	*No*	*Yes*
Ana	*No*	*Yes*	*No*	*Yes*	*No*
Sam	*Yes*	*No*	*Yes*	*No*	*No*
Hans	*Yes*	*No*	*No*	*No*	*Yes*

The size of this matrix is $m \times k$, where m is the number of users and k is the number of features identified for movies. In this case, k is 5 (because there are 5 features, "Comedy", ..., "Biopic").

So if "Yes" and "No" are replaced with 1 and 0, then the preferences of users/persons will look more like a vector that can be used in a dot product.

Persons	Comedy	Action	Documentary	Thrillers	Biopic
Dana	1	0	0	0	1
Ana	1	1	0	1	0
Sam	1	0	1	0	0
Hans	0	0	0	0	1

Now, let's imagine we somehow figure out that a movie has a few frames of comedy, a few frames of action, and so on. So each movie can be expressed as a row in this matrix:

Movie Genres	Movie1	Movie2	Movie3	Movie3	Movie5
Comedy	1	0	0	1	0
Action	0	1	0	1	0
Documentary	0	0	1	0	1
Thrillers	0	1	0	0	1
Biopic	0	0	1	0	0

This matrix has the size $k \times n$ (where n is the number of movies).

Space advantage...

To see how much space benefit this scheme brings, just assign some realistic number to all variables m, n, k.

Let's say you are doing this for some really popular platform like Amazon Prime; then the number of users m will be in the range of a million. Let's assume in a realistic scale there are 1000 movies to start with. There can be more but this is a good ballpark number. Also assume that we can distribute the preferences of users and genres of movies in some 20ish features. That makes m = 1 Million, n = 1000, and k = 20.

Now if we had stored the numbers in a big matrix number of elements, it would have been a gargantuan 1 Billion. But storing them in two separate matrices makes that number come down to only 20 Million.

Let's predict…

Movies are generally categorized to be in multiple genres. In these examples, "Movie3" is an action comedy, while "Movie5" is a thrilling documentary.

To know whether we should recommend "Movie1" for "Dana", if she had not already watched it, we needed to find out the dot product of the first row of the first matrix and the first column of the second matrix. This will be $[1, 0, 0, 0, 1] \cdot [1\,0\,0\,0\,0]$, which is 1.

Ana loves comedy and action, so for Movie5, her rating will be $[11010] \cdot [11000]$, which sums to 2.

Just to remind you, dot product is the summation of entries calculated from index-wise multiplication. So $[1, 1, 0, 1, 0] \cdot [1, 1, 0, 0, 0] = 1 \times 1 + 1 \times 1 + 0 \times 0 + 1 \times 0 + 0 \times 0 = 2$.

All these sound good, but it is too good to expect such data to be available for real-life scenarios where in most occasions the data itself must be procured in first few months/years.

Finding the right factorization of the big matrix

Finding the right set of factor matrices is an iterative process. Two matrices (one for the user-feature and another for the item-feature) are initialized with random values, and then a chosen algorithm goes back and forth many times over until the weights in these two matrices produce close-enough approximation of the true ratings provided by the big matrix.

The algorithm needs a way to measure whether it is approaching the right value or going farther from it. It does so trying to minimize the following function (also called a loss function):

$$L = \sum_{u,i \in S} \left(r_{ui} - \mathbf{x}_u^{\mathrm{T}} \cdot \mathbf{y}_i \right)^2 + \lambda_x \sum_u \left\| \mathbf{x}_u \right\|^2 + \lambda_y \sum_u \left\| \mathbf{y}_i \right\|^2$$

In this case, x represents users and y represents items. r_{ui} denotes rating for item i given by user u. So the computer tries to minimize the squared error. $\sum_{u,i \in S} \left(r_{ui} - \mathbf{x}_u^T \cdot \mathbf{y}_i \right)^2$ is the squared error. The part on the right of the equation is called regularization. This is used for preventing the system from overfitting the data.

So conceptually, the algorithm takes a set of randomized entry populated vectors for feature matrices (factor matrices) and then keeps on iterating until it hits an aggregable approximation of the true ratings (ratings provided by some user for some items).

Gradient descent family of algorithms are generally used to solve for the weights in the factor matrices.

Modifying hyperparameters in ML.NET

You can modify all these hyperparameters, a parameter whose value is used to control the learning process, for tuning the algorithm via `MatrixFactorizationTrainer.Options` class from the code.

```
// Set the training algorithm
var opt = new MatrixFactorizationTrainer.Options();
opt.LossFunction = MatrixFactorizationTrainer.LossFunctionType.
```

```
SquareLossOneClass
SquareLossRegression
```

Figure 8-1. *Showing how the Loss function can be set*

`SquareLossRegression` is the default, and `SquareLossOneClass` is used for implicit recommender systems where it is to recommend whether the user will click, buy, watch an advert, and so on.

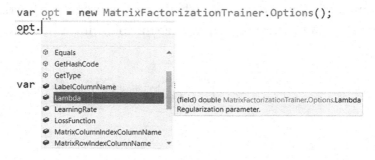

Figure 8-2. *Showing how the regularization parameter can be set*

Doing matrix factorization using Model Builder

Step 1: Select the scenario as Recommendation (Figure 8-3).

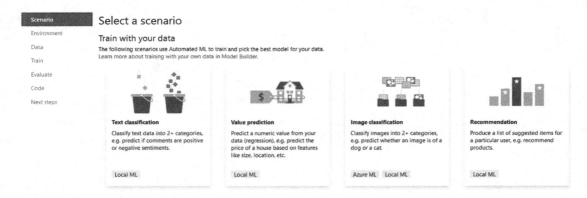

Figure 8-3. *Locating the scenario*

Step 2: Select where you want to train the model (Figure 8-4).

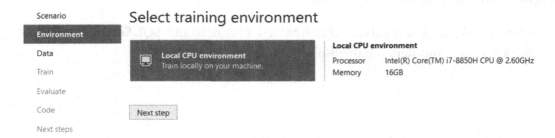

Figure 8-4. *Selecting the training environment*

At the time of this writing (August 2020), this is only possible to train it locally.

Step 3: Select the training file (Figure 8-5).

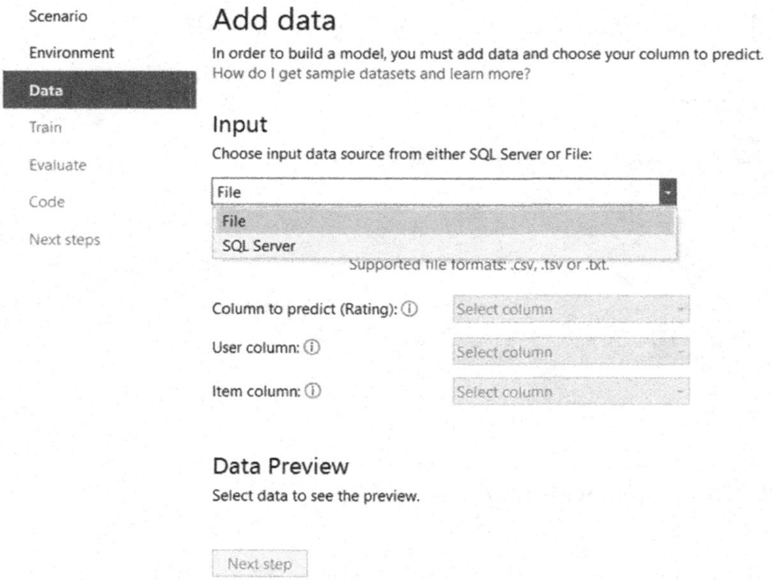

Figure 8-5. *Finding the training data*

The wizard can read the data either from a flat file or from a SQL Server database.

Step 4: Verify your data (Figure 8-6).

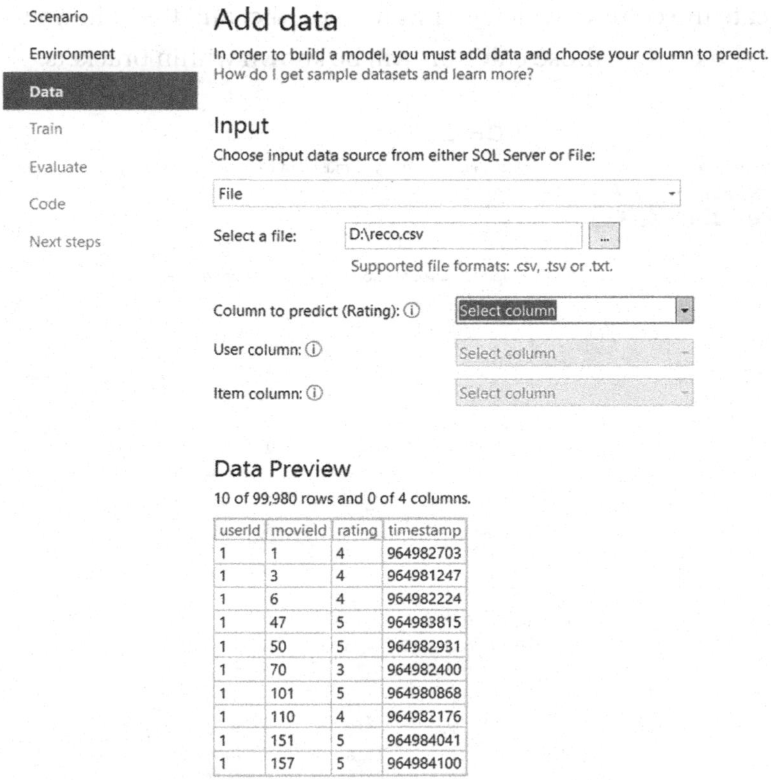

Scenario

Environment

Data

Train

Evaluate

Code

Next steps

Add data

In order to build a model, you must add data and choose your column to predict.
How do I get sample datasets and learn more?

Input

Choose input data source from either SQL Server or File:

File ▾

Select a file: | D:\reco.csv | ... |

Supported file formats: .csv, .tsv or .txt.

Column to predict (Rating): ⓘ | Select column ▾ |

User column: ⓘ | Select column ▾ |

Item column: ⓘ | Select column ▾ |

Data Preview

10 of 99,980 rows and 0 of 4 columns.

userId	movieId	rating	timestamp
1	1	4	964982703
1	3	4	964981247
1	6	4	964982224
1	47	5	964983815
1	50	5	964982931
1	70	3	964982400
1	101	5	964980868
1	110	4	964982176
1	151	5	964984041
1	157	5	964984100

Figure 8-6. *Previewing and verifying data*

139

Step 5: Locate the column which you want to predict and the column to use for training. Once you do that, those columns will be shown within brackets.

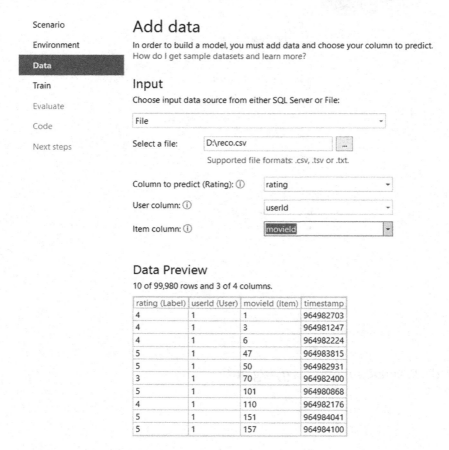

Figure 8-7. *Annotating the data for training the model*

Step 6: Leave the default time of 10 secs to train (Figure 8-8).

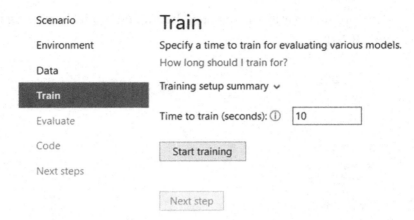

Figure 8-8. *Training the model*

Step 7: Wait for the training to finish. Once done, it will show the results like this (Figure 8-9).

Scenario

Environment

Data

Train

Evaluate

Code

Next steps

Train

Specify a time to train for evaluating various models.
How long should I train for?

Training setup summary ∨

Time to train (seconds): ⓘ 10

[Train again]   ✓ Training complete

Training results

Best quality (RSquared):   0,2823
Best model: MatrixFactorization
Training time: 8,95 seconds
Models explored (total):   11

[Next step]

Figure 8-9. *Training the model is done*

Step 8: You can optionally verify the result of the trained model by visiting the Evaluate tab (Figure 8-10).

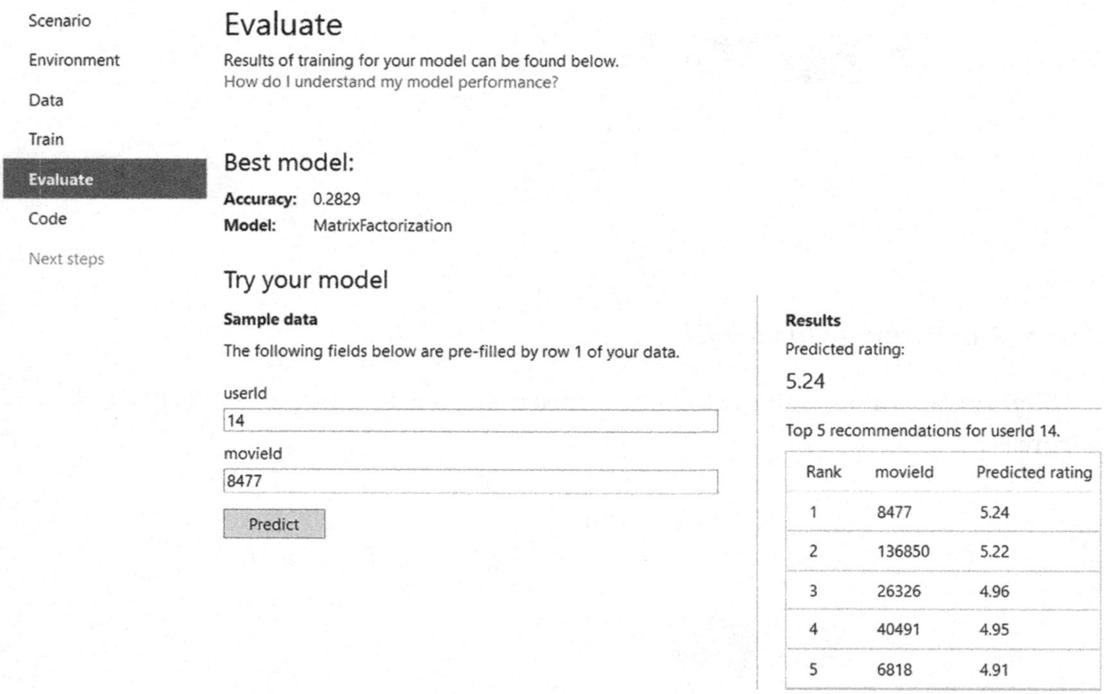

Scenario		
Environment	**Evaluate**	
Data	Results of training for your model can be found below.	
Train	How do I understand my model performance?	

Evaluate

| Code |
| Next steps |

Best model:

Accuracy: 0.2829
Model: MatrixFactorization

Try your model

Sample data

The following fields below are pre-filled by row 1 of your data.

userId

`14`

movieId

`8477`

[Predict]

Results

Predicted rating:

5.24

Top 5 recommendations for userId 14.

Rank	movieId	Predicted rating
1	8477	5.24
2	136850	5.22
3	26326	4.96
4	40491	4.95
5	6818	4.91

Figure 8-10. *Evaluation of the model*

Here, the predicted rating for user 14 for movie 8477 is shown on the right side as 5.24. It also lists the top 5 recommendations for the movie. In a real-life application though, we must limit the upper and lower predicted rating between 1 and 5. So 5.24 will be capped to 5.

Step 9: The next step is to use this generated code to the solution. Once confirmed, Model Builder adds these couple of projects to the solution.

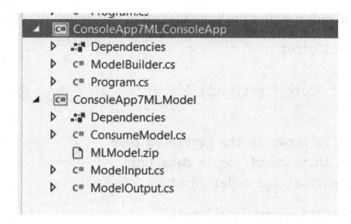

Figure 8-11. *Showing added projects in solution*

Here is the **ModelInput** class (Figure 8-12).

```csharp
8 references
public class ModelInput
{
    [ColumnName("userId"), LoadColumn(0)]
    2 references
    public float UserId { get; set; }


    [ColumnName("movieId"), LoadColumn(1)]
    2 references
    public float MovieId { get; set; }


    [ColumnName("rating"), LoadColumn(2)]
    0 references
    public float Rating { get; set; }


    [ColumnName("timestamp"), LoadColumn(3)]
    0 references
    public float Timestamp { get; set; }
}
```

Figure 8-12. *ModelInput class*

The output will only produce the score measured as R-Squared:

```
public class ModelOutput
{
    public float Score { get; set; }
}
```

This can be used as shown in the generated code.
```
// Create single instance of sample data from
//first line of dataset for model input

ModelInput sampleData = new ModelInput()

{
        UserId = 1F,
        MovieId = 1F,
};

// Make a single prediction on the sample data and print results
var predictionResult = ConsumeModel.Predict(sampleData);
```

To produce a recommended list of movies, scores for all movies have to be calculated and then sorted in descending order.

Summary

At this moment, ML.NET offers matrix factorization as the recommender system trainer. There are several other memory-based models that work well for recommendation problems, which are not available as a part of ML.NET APIs but you can import them as an ONNX models and consume in your .Net applications via ML.NET.

Recommendation can also be approached as a similarity measure problem as in information retrieval. In that approach, users are recommended movies/items that have similar attributes.

CHAPTER 9

Anomaly Detection

That doesn't look normal. Does it?

Introduction

Detecting anomalous situations early can be a lifesaver. Imagine the catastrophes saved being able to spot a manufacturing defect in a car engine before being shipped in a million cars. The manufacturer can save huge on the potential damage control for recalling all the cars, let alone the embarrassment caused.

During some stage of pregnancy, an anomaly scan is done to predict the stage of the fetus. The motive of this scan is to abort pregnancies that will cause premature or terminally ill babies.

Anomaly detection algorithms also play a huge part in fraudulent transaction detection. The computer can spot smelly/fishy transactions (if you will) from others by measuring several aspects/features of the transaction. The time of the transaction, the amount, the speed at which the login credentials were entered by the user, and so on give all the necessary clues to the anomaly detection algorithm to be able to tell chalk (the fraudulent) and cheese (the real legitimate transaction) apart.

Objective

In this chapter, you shall learn how ML.NET can be used to spot anomalies in different datasets. After reading this chapter, you should be able to write anomaly detection algorithm for your own datasets derived from your own domain models.

© Sudipta Mukherjee 2021
S. Mukherjee, *ML.NET Revealed*, https://doi.org/10.1007/978-1-4842-6543-7_9

What's an anomaly anyway?

It's the odd one out from a group of similar items. Imagine that a person has historically (for the last 5 years) spent between $1 and $2000 per transaction on her credit card. One evening the server of the bank saw an incoming debit request of $9000 on a single transaction. This is beyond the historical limits. It might very well be legit. Maybe she got extravagant and spent more than she ever did that evening, but this is enough for the anomaly detection algorithm at the bank server to attempt to flag this as an anomaly. This is because $9000 is well beyond the max she ever spent on a single transaction for a considerable time to be used to create a profile as a representative of her buying behavior.

So, in a nutshell, anomaly means anything out of the ordinary. Anything that is so obvious that it will catch the eye.

Different types of anomalies…

There are mostly three different types of anomalies:

- Point anomaly
- Contextual anomaly
- Collective anomaly

Point anomaly

When a data point is farther from all others in the input dataset, then it is probably anomalous. In this context, the point anomalies are also called outliers. Imagine that we store the area of a house and its price in a list of tuples, and the general assumption is that if the house is big, it will be expensive. However, if we spot a really big house with really cheap asking price, then that's too good to be true and represents an outlier or a point anomaly.

Contextual anomaly

Sometimes depending on the season, what seems anomalous otherwise may sound normal. For example, during the X-mas/New Year holiday season, a person from New York may spend more every day than her average spending value otherwise. This is normal, but when the same person seems to spend more even after the holidays, something might be potentially wrong. It could be that the card is stolen. As you can see, it is really hard to determine that whether we are dealing with a contextual anomaly or a point anomaly. The knowledge of the seasonal events has to be reconciled together to get to something useful.

Collective anomaly

Sometimes data points are not considered anomalous by themselves. But with the context of other points in the dataset, they can be anomalous. A missing heartbeat in an ECG, as seen in Figure 9-1, is an example of collective anomaly.

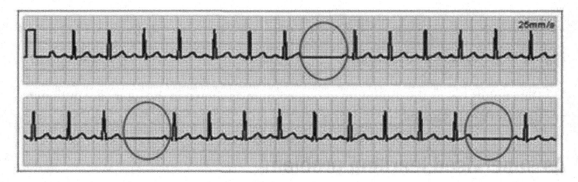

Figure 9-1. *Showing anomalous ECG*

Here, the points (highlighted by circles) themselves are not anomalous, but in the context of all points in the ECG, they are.

Different approaches to detect anomalies...

Based on how they are poised, anomaly detection problems can be represented as supervised or unsupervised/clustering problem.

As simple statistical problem

Anomaly detection can be thought of as a simple statistical problem where we need to find elements below or above the quartile range (IQR: interquartile range) as seen in Figure 9-2.

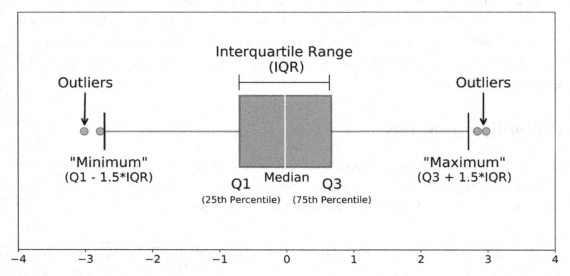

Figure 9-2. *Showing IQR range*

Data points below the minimum and maximum as shown in the figure are outliers/anomalous data instances.

As supervised learning problem

Anomaly detection can be thought of as a supervised machine learning problem if we had several labeled examples of data which are either anomalous and normal (nonanomalous). If presented like this, then several supervised machine learning classification/clustering techniques can be used to detect anomaly in the data, if there exists any.

The rationale behind supervised approaches is that it is thought that general, normal, and typical data points appear in close proximity to each other, while the anomalous instances are isolated farther apart. Because of this reason, density-based models like nearest neighbors work.

As clustering learning problem

K-Means clustering works to detect anomalies. Elements not really attached to a centroid are claimed to be anomalous. However, the challenge of anomaly detection lies in novelty. Sometimes anomalous entries appear that look nowhere near their previous incarnations. This is one of the major motivations to think of anomaly detection as unsupervised/clustering learning problem.

ML.NET offers

What ML.NET offers is various ways to locate anomalies in time series data. As shown in Figure 9-3, these are available via **Microsoft.ML.Timeseries** 1.5.1 NuGet package or beyond.

Figure 9-3. *Showing NuGet Package required*

Time series anomaly detection algorithms are implemented as extension method on TransformsCatalog. Table 9-1 shows these extension functions and their purposes.

Table 9-1. *Showing methods to detect point anomalies*

Function Name	Description
DetectChangePointBySsa	Create SsaChangePointEstimator, which predicts change points in time series using singular spectrum analysis (SSA).
DetectEntireAnomalyBySrCnn	Create Microsoft.ML.TimeSeries.SrCnnEntireAnomalyDetector, which detects time series anomalies for entire input using SRCNN algorithm.
DetectIidChangePoint	Create IidChangePointEstimator, which predicts change points in an independent identically distributed (i.i.d.) time series based on adaptive kernel density estimations and martingale scores.
DetectSpike	Create IidSpikeEstimator, which predicts spikes in independent identically distributed (i.i.d.) time series based on adaptive kernel density estimations and martingale scores. C#
DetectSpikeBySsa	Create SsaSpikeEstimator, which predicts spikes in time series using singular spectrum analysis (SSA).

SRCNN algorithm...

The goal of this algorithm is to predict a score for each input data point. More formally, if the input data is represented as $x_1, x_2, x_3, ..., x_n$, then this algorithm tries to predict the anomaly score for each point in the input data represented as $y_1, y_2, y_3, ..., y_n \in (0, 1)$ as seen in Figure 9-4.

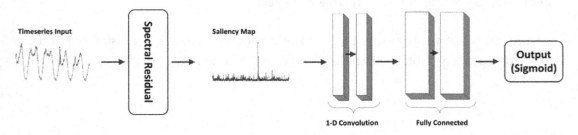

Figure 9-4. *Representation of SRCNN algorithm*

So, each point in the input either represents a time series anomaly or not. It is 1 if the point represents an anomaly and 0 otherwise.

SRCNN first uses Spectral Residual of the input data and then uses the output of this as the input of a CNN (convolutional neural net) to calculate if the point is anomalous or not.

For more pointers about the inner workings of the algorithm, follow the original post at https://techcommunity.microsoft.com/t5/ai-customer-engineering-team/overview-of-sr-cnn-algorithm-in-azure-anomaly-detector/ba-p/982798.

ML.NET encapsulation...

You can use this from ML.NET via the encapsulated method on `AnomalyDetection` transformer as shown in Figure 9-5.

```
var outputDataView = ml.AnomalyDetection.DetectEntireAnomalyBySrCnn
    (dataView,
```

```
▲ 1 of 2 ▼  (extension) Microsoft.ML.IDataView TimeSeriesCatalog.DetectEntireAnomalyBySrCnn (
              IDataView input,
              string outputColumnName,
              string inputColumnName,
              [double threshold = 0.3],
              [int batchSize = 1024],
              [double sensitivity = 99],
              [SrCnnDetectMode detectMode = SrCnnDetectMode.AnomalyOnly])
Create SrCnnEntireAnomalyDetector, which detects timeseries anomalies for entire input using SRCNN algorithm.
input: Input DataView.  F1 for help
```

Figure 9-5. *Showing tooltip on the SRCNN method in ML.NET*

Table 9-2 shows all the parameters this method takes and their purposes.

Table 9-2. *Details of the parameters of the SRCNN method*

Parameter	Purpose
OuputColumnName	Name of the column resulting from data processing of `inputColumnName`. The column data is a vector of `Double`. The length of this vector varies depending on `detectMode`.
InputColumnName	Name of column to process. The column data must be `Double`.
Threshold	The threshold to determine an anomaly. An anomaly is detected when the calculated SR raw score for a given point is more than the set threshold. This threshold must fall between [0,1], and its default value is 0.3.
batchSize	Divide the input data into batches to fit `srcnn` model. When set to -1, use the whole input to fit model instead of batch by batch; when set to a positive integer, use this number as batch size. Must be -1 or a positive integer no less than 12. Default value is 1024.
sensitivity	Sensitivity of boundaries, only useful when `srCnnDetectMode` is `AnomalyAndMargin`. Must be in [0,100]. Default value is 99.
detectMode	An enum type of `SrCnnDetectMode`. When set to `AnomalyOnly`, the output vector would be a 3-element Double vector of (`IsAnomaly`, `RawScore`, `Mag`). When set to `AnomalyAndExpectedValue`, the output vector would be a 4-element Double vector of (`IsAnomaly`, `RawScore`, `Mag`, `ExpectedValue`). When set to `AnomalyAndMargin`, the output vector would be a 7-element Double vector of (`IsAnomaly`, `AnomalyScore`, `Mag`, `ExpectedValue`, `BoundaryUnit`, `UpperBoundary`, `LowerBoundary`). The RawScore is output by SR to determine whether a point is an anomaly or not; under AnomalyAndMargin mode, when a point is an anomaly, an AnomalyScore will be calculated according to sensitivity setting. Default value is `AnomalyOnly`.

Using anomaly detection to spot spikes in sales data…

We can use the preceding method to detect spikes (which are essentially anomalies) in the input sales data. The next example will show how we can tweak several of the parameters discussed in Table 9-2 to spot anomalies.

Step 1: Create a console app in Visual Studio 2019.

```
using System;
using System.Collections.Generic;
using System.Linq;
using System.Text;
using System.Threading.Tasks;

namespace AnomalyDetect
{
    class Program
    {
        static void Main(string[] args)
        {
        }
    }
}
```

Step 2: Get the necessary NuGet packages.

Go to Tool ➤ NuGet Package Manager ➤ Package Manager Console (Figure 9-6).

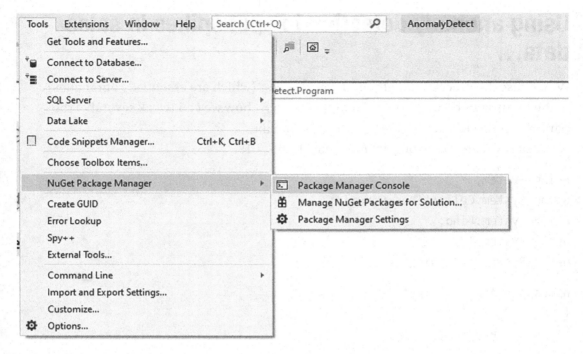

Figure 9-6. *Showing Package Manager Console menu*

When the console appears, get the two packages as seen in Figures 9-7 and 9-8.

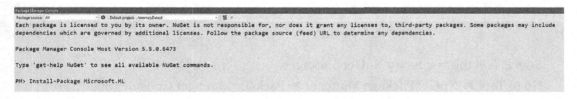

Figure 9-7. *Showing the NuGet Package is installed*

```
Package Manager Console
Package source: All          ▾ ⚙ Default project: AnomalyDetect          ▾ 延 ▪
Successfully installed 'System.Threading.Tasks.Extensions 4.5.4' to AnomalyDetect
Adding package 'System.Threading.Channels.4.7.1' to folder 'C:\Users\Sudipta\source\repos\AnomalyDetect\packages'
Added package 'System.Threading.Channels.4.7.1' to folder 'C:\Users\Sudipta\source\repos\AnomalyDetect\packages'
Added package 'System.Threading.Channels.4.7.1' to 'packages.config'
Successfully installed 'System.Threading.Channels 4.7.1' to AnomalyDetect
Adding package 'Microsoft.ML.1.5.1' to folder 'C:\Users\Sudipta\source\repos\AnomalyDetect\packages'
Added package 'Microsoft.ML.1.5.1' to folder 'C:\Users\Sudipta\source\repos\AnomalyDetect\packages'
Added package 'Microsoft.ML.1.5.1' to 'packages.config'
Successfully installed 'Microsoft.ML 1.5.1' to AnomalyDetect
Executing nuget actions took 6.05 sec
Time Elapsed: 00:00:21.4897167
PM> Install-Package Microsoft.ML.Timeseries
```

Figure 9-8. *Showing that ML.Timeseries NuGet package is also installed*

Once successful, these references will be shown on the References of the project as shown in Figure 9-9.

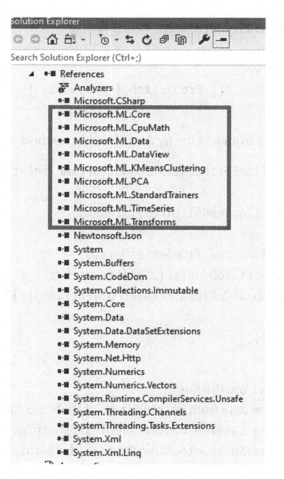

Figure 9-9. *Showing all these dlls in the References*

At this stage, you are ready to use the data.

Step 3: Add the necessary using directive.

```
using Microsoft.ML;
```

Step 4: Create an ML context.

```
MLContext ml = new MLContext();
```

Step 5: Add the following classes to the solution:

```
private class TimeSeriesData
{
    public double Value { get; set; }
}

private class SrCnnAnomalyDetection
{
            [VectorType]
            public double[] Prediction { get; set; }
}
```

Step 6: Load the data from text file by the following method:

```
private static List<TimeSeriesData> LoadDataFromFile(string fileName)
{
            return File.ReadAllLines(fileName)
            .Skip(1)
            .Select(f => new TimeSeriesData()
        { Value = Convert.ToDouble(f.Split(new char[] { ',' },
            StringSplitOptions.RemoveEmptyEntries)[1])
        })
            .ToList();
}
```

Step 7: Load the data from the file.

You can download the data from https://raw.githubusercontent.com/
dotnet/machinelearning-samples/master/samples/csharp/getting-started/
AnomalyDetection_Sales/SpikeDetection/Data/product-sales.csv.

After that from the Main method, you can load this into a list:

```
var data = LoadDataFromFile(@"D:\product-sales.csv");
```

Step 8: Convert the data to an IDataView instance.

```
var dataView = ml.Data.LoadFromEnumerable(data);
```

Step 9: Prepare the input and output column.

```
string outputColumnName = nameof(SrCnnAnomalyDetection.Prediction);
string inputColumnName = nameof(TimeSeriesData.Value);
```

Step 10: Perform the batch anomaly detection for each input data point.

```
// Do batch anomaly detection
var outputDataView = ml.AnomalyDetection.DetectEntireAnomalyBySrCnn
  (dataView,
   outputColumnName,
   inputColumnName,
   threshold: 0.30,
   batchSize: -1,
   sensitivity: 91,
   detectMode: SrCnnDetectMode.AnomalyAndExpectedValue);
```

Step 11: Get the newly created column.

```
var predictionColumn = ml.Data.CreateEnumerable<SrCnnAnomalyDetection>(
                outputDataView, reuseRowObject: false);
```

Step 12: Loop through the predicted column to find the spikes.

```
  foreach (var prediction in predictionColumn)
  {
            if(prediction.Prediction[2]>0.3)
     {
            Console.WriteLine($"Detected spike at {data[k].Value}");
     }
    k++;
  }
```

This prints the following. And obviously, the first entry has to be ignored as this is the beginning of the data values.

```
Detected spike at 271
Detected spike at 150.9
Detected spike at 341.5
Detected spike at 426.6
Detected spike at 687
```

Summary

ML.NET offers anomaly detection for time series analysis, but as mentioned in the beginning of the chapter, anomalies can occur in any data. Hopefully, other statistical methods like IQR and related methods will be incorporated in the framework.

CHAPTER 10

Object Detection

Can you spot the cat in the photo?

Introduction

Automatically detecting objects in an image either static or derived from a constant video capturing source has numerous applications as you can imagine. Here are just a very few of them:

Improved photo search capability

We all search for photographs by things like phrases. Sometimes those phrases can be used to look up already tagged photos from the Internet. However sometimes nothing is available, and search engines go find them from frames of videos or from photo archives using image search algorithms that employ some kind of object detection and classification. And if these returned images are useful to the end user, they get tagged with the search phrases accordingly so that the future lookups become faster.

Video surveillance (real-time object detection)

With modern really fast algorithms like YOLO and its variations, it is now very easy and fast (more importantly) to run an object detection on real-time video frames. I encourage you to see the video on YOLO site: `https://pjreddie.com/darknet/yolov2/`.

Object counting

It is often required to find out an approximate count (within a tolerable loss/gain of percentage) of objects present in a photo. One example is to count the number of heads; this can be helpful in estimating crowd density automatically in an event to find out popularity or measuring the social success of an event. Another example could be automatic medical diagnosis. Another example of object counting is the process to improve the number counting learning experience for toddlers. Imagine a photo with some bottles and a toddler is asked to count the bottles. If the count given by the kid matches with that of the model, then the kid gets a score and a suggestion or a hint otherwise. This kind of self-teaching capabilities can be easily built and can be personalized for better reach.

Automatic captioning of photographs

Imagine a photo with multiple bikes, few trains, and lots of people; this photo can be captioned "At the busy station".

Objective

ML.NET offers the capability to run pretrained models from ONNX and TensorFlow that allow to detect objects from photos/images. Since object detection is a very computation-heavy activity to train the models on, you shall require lot of data and time to train a model. However, consuming pretrained model for detecting several objects from your images can be simple. In this chapter, you shall see how you can use YOLO from ONNX model zoo to detect objects in images. I hope the chapter will leave you with enough inspiration and knowledge to use other models from ONNX model zoo.

How YOLO works

YOLO stands for You Only Look Once (YOLO). The algorithm takes an input image (known as "image" and is represented by $3 \times 416 \times 416$ tensor). The output of the algorithm is a tensor with $125 \times 13 \times 13$ dimension and is called "grid".

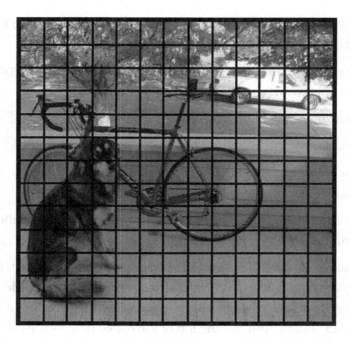

Figure 10-1. *Showing how YOLO splits the input image into 13 × 13 cells*

YOLO splits the given image into 13 × 13 (or 169 cells). Each cell produces or is bound to produce five bounding boxes. Each bounding box is represented by 25 variables.

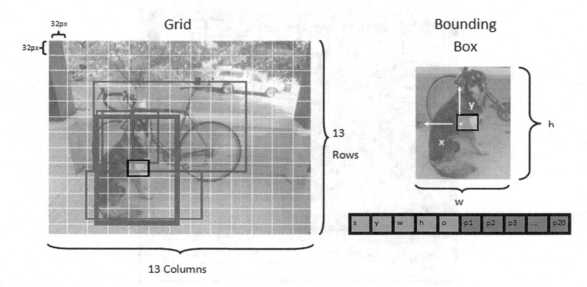

Figure 10-2. *(taken from ML.NET documentation from Microsoft)*

- x is the x position of the bounding box center relative to the grid cell it's associated with.

- y is the y position of the bounding box center relative to the grid cell it's associated with.

- w is the width of the bounding box.

- h is the height of the bounding box.

- o is the confidence value that an object exists within the bounding box, also known as *objectness* score.

- p1-p20 is the class probabilities for each of the 20 classes predicted by the model.

A bounding box is the area of interest in which an object is detected. Each bounding box gets a probability distribution of 20 values that represent the confidence score for each class/type. By default, the algorithm ignores anything that has a confidence score of less than 0.3 or 30%. And it is the job of the callee (you as the caller of YOLO) to determine the bounding boxes and types.

Predicted bounding boxes might look like this. The fatter the bounding box, the higher is the confidence that some predefined object is there in that area.

Figure 10-3. *Showing bounding boxes*

The next step after obtaining the bounding box details is to sort the bounding boxes by a calculated score by somehow gluing together the confidence score (objectness score) and the probability distribution score. For example, the following image shows that the YOLO algorithm is almost certain that the left-bottom bounding box colored "yellow" has the object "dog" in it (Figure 10-4).

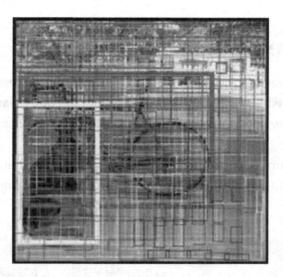

Figure 10-4. *Showing all bounding boxes*

From this obviously, the top three bounding boxes (that have the fattest boundaries) stand out (Figure 10-5).

Figure 10-5. *Showing the final result of a YOLO model*

There are 169 cells and for each cell there are 5 bounding boxes so there are in total 845 bounding boxes. Most of these bounding boxes have a very low objectness score (or confidence score if you will). But the neural network saw and predicted results for each of these bounding boxes together and at once. That's why the name is YOLO (You only look once!).

Removing overlapping boxes…

As you can see from the example earlier, there will be so many overlapping boxes in the prediction. But obviously, we need a way to cancel out the ones that are not as good. The algorithm to remove overlapping bounding boxes is "Non-maximum Suppression" (a.k.a. NMS).

As they say, a picture is worth a hundred words, so here is what NMS does to a bunch of overlapping bounding boxes. It eliminates the overlapping bounding boxes with lesser confidences.

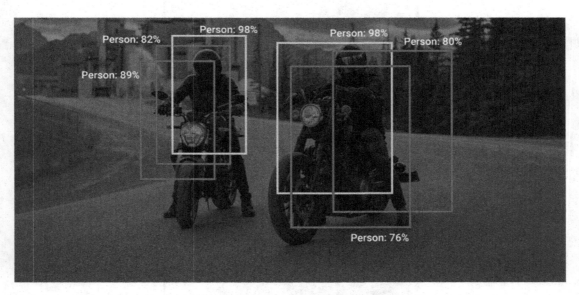

Figure 10-6. *What NMS does to overlapping bounding boxes*   .

Non-maximum-suppression algorithm steps

Step 1

First, the bounding box with the highest confidence score is selected, and it is added to the final list of proposed bounding boxes. And this box is removed from the list of suggestion provided by YOLO.

Step 2

For this current bounding box, calculate the IOU (intersection over union) with all the other proposed bounding boxes by YOLO. And if the calculated IOU is greater than the threshold, then the other bounding boxes need to be removed from the set of boxes.

Step 3

Pick the next bounding box with the highest confidence score and continue step 2 until all the bounding boxes are touched or removed. At the end, you shall be left with only those bounding boxes that are of interest.

What's IOU of two bounding boxes?

IOU is the ratio of area of intersection and union of two bounding boxes. Figure 10-7 shows this visually.

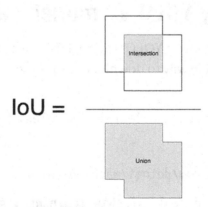

Figure 10-7. *Showing the ratio of intersection and union visually*

NMS pseudocode…

Here, B means the set of bounding boxes, c is the confidence score threshold, and λ_{nms} is the threshold for overlap.

Non-Max Suppression

---
procedure $NMS(B,c)$
    $B_{nms} \leftarrow \emptyset$
    for $b_i \in B$ **do**
        $discard \leftarrow \text{False}$
        for $b_j \in B$ **do**
            if $\text{same}(b_i, b_j) > \lambda_{\mathbf{nms}}$ **then**
                if $\text{score}(c, b_j) > \text{score}(c, b_i)$ **then**
                    $discard \leftarrow \text{True}$
        if not $discard$ **then**
            $B_{nms} \leftarrow B_{nms} \cup b_i$
    return B_{nms}

Figure 10-8. *NMS pseudocode*

Consume the tiny YOLO V2 model via ML.NET

The easiest way to experiment with YOLO using ML.NET is to download the samples from GitHub. The best possible way to download is to clone the repo.

Go to

https://github.com/dotnet/machinelearning-samples.

Clone it via git as

git clone https://github.com/dotnet/machinelearning-samples.git

Once you have it on disk, go to the **\machinelearning-samples\samples\csharp\ end-to-end-apps\ObjectDetection-Onnx** folder and open the solution using Visual Studio 2019.

Once the project loads, try to build it. Ensure that you are connected to the Internet because it will have to restore several NuGet packages.

Once everything runs smoothly, you should expect to see the solution explorer like this (Figure 10-9).

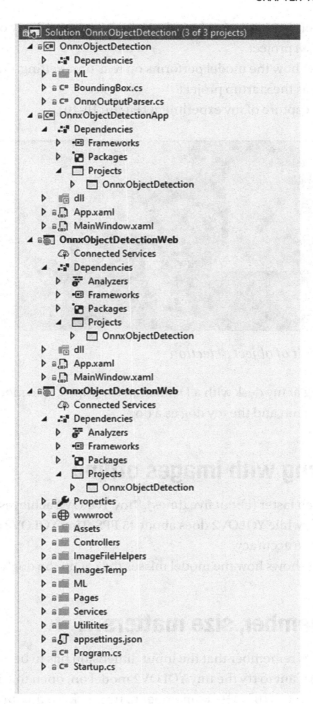

Figure 10-9. *What the solution will look like*

Notice that two end-to-end apps (one desktop and one web) rely on the **OnnxObjectDetection** project.

If you want to see how the model performs on real-time setting, choose the OnnxDetectionApp as the startup project.

Here is a screen capture of my experiment (Figure 10-10).

Figure 10-10. *Result of object detection*

Here, I am sitting at my desk with a Dalmatian toy puppy. The model correctly identifies me as a person and the toy dog as a dog.

Experimenting with images offline

Tiny YOLOV2 is much faster (about five times). Tiny YOLOV2 achieves up **to 244 FPS** (frames per second), while YOLOV2 does about 45 FPS. Tiny YOLOV2 achieves this by sacrificing some of the accuracy.

This experiment shows how the model misses the cat on the dog's head.

Always remember, size matters...

To use images, always remember that the input dimension has to be 416 × 416. So whatever image you want to try the tiny YOLOV2 model on, open that in MS Paint and change the dimension to 416 × 416. Remember to uncheck "Maintain aspect ratio".

Figure 10-11. *Setting the input dimensions*

Now, select the **OnnxObjectDetectionWeb** as the startup project. It brings up a page where you can upload your image. And as soon as the upload finishes, the model takes over and draws the bounding boxes along with confidence scores. For my cat and dog picture, it totally misses the cat on top of the dog's head.

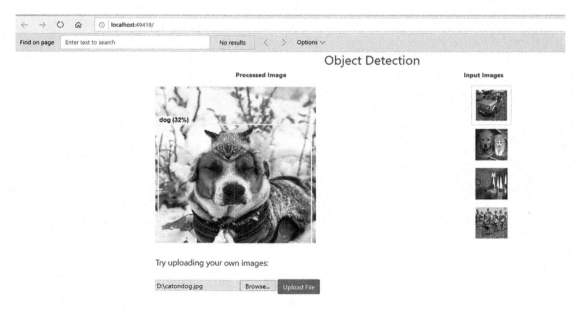

Figure 10-12. *Example of uploading image for object detection*

Experimenting with a different model…

If you want to use a different model, you can simply find the model in ONNX Model Zoo (`https://github.com/onnx/models`) and remember to replace the one in the ONNXModels folder.

Summary

ML.NET offers features to use pretrained ONNX and TensorFlow models, so you can easily experiment with several deep learning models for numerous different types of machine learning activity. I hope, in the future, ML.NET will also provide capabilities to train a model and transform it to an ONNX model.

Index

Printed in the United States
By Bookmasters